D1000602

RESEARCH AND PERSPECTIVES IN NEUROSCIENCES

Fondation Ipsen

Editor

Yves Christen, Fondation Ipsen, Paris (France)

Editorial Board

Albert Aguayo, McGill University, Montreal (Canada)
Philippe Ascher, Ecole Normale Supérieure, Paris (France)
Alain Berthoz, Collège de France, CNRS UPR 2, Paris (France)
Jean-Marie Besson, INSERM U 161, Paris (France)
Emilia Bizzi, Massachusetts Institute of Technology, Boston (USA)
Anders Bjorklund, University of Lund (Sweden)
Floyd Bloom, Scripps Clinic and Research Foundation, La Jolla (USA)
Joël Bockaert,Centre CNRS-INSERM de Pharmacologie Endocrinologie,
   Montpellier (France)
Pierre Buser, Institut des Neurosciences, Paris (France)
Jean-Pierre Changeux, Collège de France, Institut Pasteur, Paris (France)
Carl Cotman, University of California, Irvine (USA)
Steven Dunnett, University of Cambridge, Cambridge (UK)
George Fink, Medical Research Council, Edingburgh (UK)
Fred Gage, Salk Institute, La Jolla (USA)
Jacques Glowinski, Collège de France, Paris (France)
Claude Kordon, INSERM U 159, Paris (France)
Michel Lacour, CNRS URA 372, Marseille (France)
Michel Le Moal, INSERM U 259, Bordeaux (France)
Gary Lynch, University of California, Irvine (USA)
Brenda Milner, McGill University, Montreal (Canada)
John Olney, Washington University Medical School, Saint Louis (USA)
Alain Privat, INSERM U 336, Montpellier (France)
Alien Roses, Duke University Medical Center, Durham (USA)
Constantino Sotelo, INSERM U 106, Paris (France)
Jean-Didier Vincent, Institut Alfred Fessard, CNRS, Gif-sur-Yvette (France)
Bruno Will, Centre de Neurochimie du CNRS/INSERM U 44,
Strasbourg (France)

B. Bontempi   A. Silva
Y. Christen (Eds.)

QP
406
M43
2007
Van

Memories: Molecules and Circuits

With 61 Figures

Bontempi, Bruno, Ph.D.
Laboratoire de Neurosciences Cognitives
CNRS UMR 5106
Université de Bordeaux 1
Avenue des Facultés
33405 Talence
France
e-mail: b.bontempi@lnc.u-bordeaux1.fr

Silva, Alcino J., Prof. Dr.
Departments of Neurobiology
Psychiatry and Biobehavioral Sciences
Psychology and Brain Research Institute
695 Young Drive South
Room 2357 Box 951761, UCLA
Los Angeles, CA 90095-1761
USA
e-mail: silvaa@mednet.ucla.edu

Christen, Yves, Ph.D.
Fondation IPSEN
Pour la Recherche Thérapeutique
24, rue Erlanger
75781 Paris Cedex 16
France
e-mail: yves.christen@ipsen.com

ISBN 978-3-540-45698-8 Springer Berlin Heidelberg New York

Cataloging-in-Publication Data applied for Bibliographic information published by Die Deutsche Bibliothek
Die Deutsche Bibliothek lists this publication in the Deutsche Nationalbibliografie; detailed biblio-graphic data is available in the Internet at <http://dnb.ddb.de>.

This work is subject to copyright. All rights are reserved, whether the whole or part of the material is concerned, specifically the rights of translation, reprinting, reuse of illustrations, recitation, broadcasting, reproduction on microfilm or in any other way, and storage in data banks. Duplication of this publication or parts thereof is permitted only under the provisions of the German Copyright Law of September 9, 1965, in its current version, and permissions for use must always be obtained from Springer. Violations are liable for prosecution under the German Copyright Law.

Springer is a part of Springer Science+Business Media
springer.com

© Springer-Verlag Berlin Heidelberg 2007

The use of general descriptive names, registered names, trademarks, etc. in this publication does not imply, even in the absence of a specific statement, that such names are exempt from the relevant protective laws and regulations and therefore free for general use.

Product liability: The publishers cannot guarantee the accuracy of any information about dosage and application contained in this book. In every individual case the user must check such information by consulting the relevant literature.

Cover design: *design & production*, Heidelberg, Germany
Typesetting and production: LE-TeX Jelonek, Schmidt & Vöckler GbR, Leipzig, Germany
Printed on acid-free paper   27/3100/YL   5 4 3 2 1 0
SPIN 11867869

Preface

Memory can be typically defined as the brain function enabling the encoding, storage and retrieval of sensory information. In operational terms, this definition implies that our central nervous system not only processes various sensory modalities, be they visual, tactile, auditory, olfactory or gustatory, but is also capable of forming, organizing and conserving memory traces for extended periods of time. At both psychological and physiological levels, there is now a consensus that memory must no longer be seen as a unitary phenomenon but rather as an ensemble of dynamic processes, each one being subserved by different brain regions organized into multiple memory systems that support different forms of memory and that constantly interact to ensure optimal performance during any given cognitive challenge.

Despite remarkable progress achieved over the last 30 years in our understanding of the neural bases of cognitive processes and associated pathologies, the questions of how, where and when memory traces are formed in the brain remain central issues and continue to fuel much debate in the field of cognitive neuroscience. How do neuronal systems encode, consolidate and retrieve memory? How are memories embedded into complex neuronal networks? How do molecular mechanisms modulate the neuronal plasticity and functioning of these networks during memory processing? What are the fundamental units of computation in the brain? Thanks to the development of novel approaches, including transgenic techniques, functional brain imaging, multiple cell recording, functional genomics and proteomics, the last decade has witnessed dramatic advances in the neurobiology, neuroanatomy, neurophysiology and neuropathology of learning and memory processes. The main aim of the *Colloque Médecine et Recherche*, titled "Memories: Molecules and Circuits," organized by the *Fondation Ipsen* in Paris on April 24, 2006, and upon which the present set of chapters is based, was to survey these recent advances and to provide an integrative view of molecular, cellular, and systems level mechanisms underlying cognitive processes in animals, nonhuman primates and humans. During the conference, current state of the art and future avenues were discussed by distinguished speakers who provided not only an overview of the underlying neurobiology of cognitive processes from a basic science standpoint but also focused on clinical and therapeutic aspects surrounding impairments associated with disorders that affect cognition. Views from these different speakers are reflected here in the form of eleven chapters that tackle key issues in the field of cognitive neuroscience.

Studies of brain-damaged patients have provided remarkable insights into the neuroanatomy of memory and into how memories are organized and stored in the brain. One chapter illustrates this line of studies and focuses on the role of the medial temporal lobe, including the hippocampus, with respect to different memory forms such as declarative memory (i.e., conscious memory for facts and events) and non declarative memory (i.e., unconscious memory for skills and habits that is independent of the medial temporal lobe). Interestingly, the data reviewed point to a time-limited role for the hippocampal region in remote memory storage. While the medial temporal lobe is crucial for acquiring new memories, it does not appear to be required to recollect detailed autobiographical memories acquired long ago, this function being progressively subsumed by extrahippocampal structures, namely widespread cortical regions, as memories mature over time. Converging evidence has also been found in experimental animals, as detailed in another chapter that focuses on the dynamics of hippocampal-cortical interactions during the course of the memory consolidation process that are required for the formation of enduring memories. Functional brain imaging studies alongside pharmacological approaches and traditional lesion techniques have provided new breakthroughs in the identification of the brain regions underlying the organization of recent and remote memories. While the neuronal mechanisms driving memory consolidation continue to be hotly debated, it is now acknowledged that the formation of long-lasting memories is no longer to be seen as a simple passive phenomenon but rather as a dynamic process of continuous neuronal reorganization as new memories are being processed and subsequently stored.

How can we unravel the cellular and molecular mechanisms underlying the formation of enduring memories? Performing genetic manipulations to control gene expression is certainly one powerful approach to dissect the plethora of signaling pathways that lead to memory formation. One chapter illustrates this type of approach and provides convincing evidence that *Drosophila* constitutes a valid and integrative insect model to study some of the cellular mechanisms involved in normal or pathological human memory. As highlighted in another chapter centered on the molecular bases of remote memory in the mouse, experimental studies have started to successfully identify specific mechanisms and underlying molecules that govern structural synaptic plasticity at both the hippocampal and cortical levels, as memories progressively gain stability.

What happens when remotely acquired memories are reactivated? Memories may never be stored in a definitive form but continuously reorganized and restabilized each time they are retrieved. In this respect, the very term "consolidation" may be misleading. Indeed, upon reactivation, cortical memories have been shown to become labile again and to require another round of protein synthesis to regain stability. Two chapters tackle this critical issue and provide exciting evidence that disruption of the so-called process of "memory reconsolidation" may be a valid therapeutic strategy for treating maladaptive memory disorders such as post-traumatic stress and drug addiction.

How are memories encoded and stored? Such processes necessarily involve the constitutive units of the brain, namely neurons, which are organized into complex, interconnected cell assemblies. Plasticity phenomena in the form of changes in synaptic efficacy within existing synapses or synthesis of novel synapses can readily occur within these cell assemblies. Significant advances have been achieved in the search of the neuronal code of memories using large-scale electrophysiological techniques that allow for the recording of multiple neuronal ensembles (up to 260 neurons) simultaneously in the freely moving rodent or nonhuman primate confronted with various cognitive challenges. Three chapters in this book provide representative examples of this kind of approach which, from a conceptual point of view, highlight the high selectivity, hierarchy and cooperation between hippocampal and cortical neurons in identifying specific stimuli, analyzing the outcome (success or failure) of the behavioral response, and converting daily experiences into various memory forms. Highly integrative properties of hippocampal-cortical cell assemblies that can act as basic coding units provide a network mechanism for the brain to achieve large memory storage capacity and higher cognitive functions, such as association, abstraction or generalization, as well as the representation of space in terms of position, direction and velocity.

How is our brain capable of coordinating thoughts and action in relation to internal goals? As reviewed in another chapter, one specific region in the brain, the prefrontal cortex, seems to play a privileged role in exerting this type of cognitive control. This region appears to subserve higher-order knowledge and is particularly well suited for acquiring abstract rules and concepts needed to guide complex, goal-directed behaviors of the kind observed in nonhuman primates and humans.

The growing knowledge of the cellular and molecular mechanisms underlying memory function also has opened the door to new therapeutic strategies for the treatment of neuropsychiatric disorders. One last chapter nicely illustrates such novel strategies by pointing to new molecules that disrupt memory circuits in Alzheimer's disease and by identifying novel targets for alleviating the cognitive deficits associated with this neurodegenerative disease.

Overall, we hope that the different chapters in this book will offer a comprehensive and up-to-date overview of the current knowledge about the neuroanatomy of memory and its underlying cellular and molecular mechanisms. Clearly, mechanisms that govern memory processes are incredibly complex and a vast array of questions still remains to be addressed. However, the set of chapters presented here also demonstrate that the field of memory research is moving forward at a rapid pace. By increasingly relying on integrative approaches that involve cross-species analyses and a combination of multiple levels of analyses, ranging from innovative cognitive paradigms to state-of-the-art cellular and molecular techniques, there is no doubt that ongoing and future studies will continue to be successful in shedding new light on this fascinating field of research.

Finally, we would like to take this opportunity to thank Jacqueline Mervaillie, Sandra Marchand and the other contributing members of the *Fondation Ipsen* for their outstanding skills in the organization of this *Colloque Médecine et Recherche*. We are also grateful to Mary Lynn Gage for her excellent editorial assistance during the preparation of this book.

Bruno BONTEMPI
Alcino J. SILVA
Yves CHRISTEN

Table of Contents

Contributors

Bayley, Peter J.
VA Medical Center (116A), 3350 La Jolla Village Drive, San Diego, CA 92161, USA

Bontempi, Bruno
Laboratoire de Neurosciences Cognitives, CNRS UMR 5106, Université de Bordeaux 1, Avenue des Facultés, 33405 Talence, France

Brown, Robert A.M.
Departments of Neurobiology, Psychiatry and Biobehavioral Sciences, Psychology and Brain Research Institute, 695 Young Drive South, Room 2357 Box 951761, UCLA Los Angeles, CA 90095-1761, USA

Comas, Daniel
Gènes et Dynamique des Systèmes de Mémoire, CNRS UMR 7637, École Supérieure de Physique et Chimie Industrielle, 10 Rue Vauquelin, 75005 Paris, France

De Felice, Fernanda G.
Cognitive Neurology & Alzheimer's Disease Center, Northwestern University Institute for Neuroscience, 633 Clark Street, Evanston, IL 60208, USA

Dudai, Yadin
Department of Neurobiology, The Weizmann Institute of Science, Rehovot 76100, Israel

Durkin, Thomas P.
Laboratoire de Neurosciences Cognitives, CNRS UMR 5106, Université de Bordeaux 1, Avenue des Facultés, 33405 Talence, France

Everitt, Barry J.
Department of Experimental Psychology, MRC/Wellcome Trust Behavioural and Clinical Neuroscience Institute, University of Cambridge, Downing Street, Cambridge CB2 3EB, UK.

Ferreira, Sergio T.
Cognitive Neurology & Alzheimer's Disease Center, Northwestern University Institute for Neuroscience, 633 Clark Street, Evanston, IL 60208, USA

Klein, William L.
Cognitive Neurology & Alzheimer's Disease Center, Northwestern University Institute for Neuroscience, 633 Clark Street, Evanston, IL 60208, USA

Isabel, Guillaume
Gènes et Dynamique des Systèmes de Mémoire CNRS UMR 7637, École Supérieure de Physique et Chimie Industrielle, 10 Rue Vauquelin, 75005 Paris, France

Lacor, Pascale N.
Cognitive Neurology & Alzheimer's Disease Center, Northwestern University Institute for Neuroscience, 633 Clark Street, Evanston, IL 60208, USA

Lee, Jonathan L. C.
Department of Experimental Psychology, MRC/Wellcome Trust Behavioural and Clinical Neuroscience Institute, University of Cambridge, Downing Street, Cambridge CB2 3EB, UK.

Miller, Earl K.
The Picower Institute for Learning and Memory, RIKEN-MIT Neuroscience Research Center, and Department of Brain and Cognitive Sciences, Massachusetts Institute of Technology, 77 Massachusetts Avenue 46-6241, Cambridge, MA 02139, USA

Moser, Edvard I.
Centre for the Biology of Memory, Norwegian University of Science and Technology, NO-7489 Trondheim, Norway

Preat, Thomas
Gènes et Dynamique des Systèmes de Mémoire, CNRS UMR 7637, École Supérieure de Physique et Chimie Industrielle, 10 Rue Vauquelin, 75005 Paris, France

Sargolini, Francesca
Centre for the Biology of Memory, Norwegian University of Science and Technology, NO-7489 Trondheim, Norway

Silva, Alcino J.
Departments of Neurobiology, Psychiatry and Biobehavioral Sciences, Psychology and Brain Research Institute, 695 Young Drive South, Room 2357 Box 951761, UCLA Los Angeles, CA 90095-1761, USA

Squire, Larry R.
VA Medical Center (116A), 3350 La Jolla Village Drive, San Diego, CA 92161, USA

Suzuki, Wendy A.
Center for Neural Science, New York University, Building Meyer, Room 809, 4 Washington Place, New York, NY 10003, USA

Talton, Lynn E.
Departments of Neurobiology, Psychiatry and Biobehavioral Sciences, Psychology and Brain Research Institute, 695 Young Drive South, Room 2357 Box 951761, UCLA Los Angeles, CA 90095-1761, USA

Tsien, Joe Z.
Center for Systems Neurobiology, Departments of Pharmacology and Biomedical Engineering, Boston University, L-601, 715 Albany Street, Boston, MA 02118-2526, USA

Wiltgen, Brian J.
Departments of Neurobiology, Psychiatry and Biobehavioral Sciences, Psychology and Brain Research Institute, 695 Young Drive South, Room 2357 Box 951761, UCLA Los Angeles, CA 90095-1761, USA

The Neuroanatomy and Neuropsychology of Declarative and Nondeclarative Memory

Peter J. Bayley[2] and *Larry R. Squire*[1234]

The hippocampus and anatomically related structures in the medial temporal lobe support the capacity for declarative memory (Squire and Zola-Morgan 1991; Eichenbaum and Cohen 2001). Declarative memory refers to the capacity to recollect facts and events and has the defining features of being consciously accessible and flexible in its expression. Declarative memory can be contrasted to a collection of nondeclarative forms of memory that includes skills and habits, priming, and simple forms of conditioning. These forms of memory are supported by brain structures outside the medial temporal lobe. Nondeclarative memory is expressed through performance rather than recollection and what is learned is nonconscious and dependent on the original learning conditions for full expression.

This chapter describes recent studies that illuminate the function of the medial temporal lobe and the relationship between declarative and nondeclarative memory. Three findings will be emphasized: 1) the capacity for learning new factual information (semantic memory) is always dependent on the medial temporal lobe; 2) remote memory for autobiographical events (episodic memory) is independent of the medial temporal lobe; and 3) humans have a robust capacity for habit memory that can operate independently of declarative memory and independently of the medial temporal lobe.

Failure to acquire new semantic knowledge in patients with large medial temporal lobe lesions

Semantic memory refers to the factual information that we acquire about the world. It has been observed that the ability to acquire new semantic memory is impaired following medial temporal lobe damage. Thus, patients with damage limited to the hippocampal region have moderate difficulty in acquiring new semantic memory, whereas patients with damage extending beyond the hippocampus into adjacent medial temporal lobe structures have a more severe impairment. From these observations, one could conclude that the ability to acquire new semantic knowledge is fully dependent upon the medial temporal lobe and that complete lesions to this region should abolish new semantic learning. However, a second possibility is that some semantic learning can

[1] Veterans Affairs Healthcare System, San Diego, CA 92161, USA
[2] Department of Psychiatry, University of California, San Diego, La Jolla CA 92093, USA
[3] Department of Neurosciences, University of California, San Diego, La Jolla CA 92093, USA
[4] Department of Psychology, University of California, San Diego, La Jolla CA 92093, USA

Address to correspondence: Larry R. Squire, VA Medical Center (116A), 3350 La Jolla Village Drive, San Diego, CA 92161, USA
lsquire@ucsd.edu and pbayley@ucsd.edu

Bontempi et al.
Memories: Molecules and Circuits
© Springer-Verlag Berlin Heidelberg 2007

proceed independently of the medial temporal lobe. Specifically, it has been suggested that new semantic information could be acquired directly by the neocortex (Tulving 1991; McClelland et al. 1995). Although the specific circumstances under which this cortical learning might occur have not been defined, it is thought that cortical learning should be slow and that considerable repetition and an extended period of learning would be required. It has been difficult to decide between these two possibilities because most amnesic patients available for study have some remaining medial temporal lobe tissue that may be sufficient to support new semantic learning.

We have had the opportunity to address this question by studying two patients (E.P. and G.P.), both of whom are severely amnesic as a result of nearly complete medial temporal lobe lesions bilaterally due to herpes simplex encephalitis (Bayley and Squire 2005). Both patients have extensive, virtually complete bilateral damage to the hippocampus, amygdala, entorhinal cortex, and perirhinal cortex, as well as approximately 70% of the parahippocampal cortex. The damage also extends beyond the medial temporal lobe in both patients to involve portions of the anterior insular cortex and the anterior fusiform gyrus. Both patients perform at chance on formal tests of new learning ability, for example, recall and recognition of word lists, stories, and diagrams (Stefanacci et al. 2000; Levy et al. 2004). Despite their profound memory impairments, both patients score in the normal range on tests of general intelligence and also perform normally on difficult tests of visual perceptual discrimination, including tests of the ability to discriminate among feature-ambiguous stimuli (Stark and Squire 2000; Shrager et al. 2006). If the capacity for new semantic learning is fully dependent on the medial temporal lobe, then patients E.P. and G.P. should not have acquired new semantic knowledge since the onset of their amnesia in 1992 and 1987, respectively. Alternatively, if some learning of facts can occur directly in neocortex, then E.P. and G.P. should demonstrate evidence of having acquired some new knowledge.

Five tests were given to assess what new knowledge might have been acquired since the onset of amnesia. In the first test, the patients and eight controls were asked 20 questions that were constructed to be so easy that controls would achieve near-perfect scores. Some of the questions related to specific events (e.g., "What crime was O.J. Simpson charged with?"), and other questions related to broader semantic knowledge (e.g., "What is the Internet?" and "What is the new European currency called?"). Testing occurred first in a free-recall format and then in a three-alternative, multiple-choice format. The results are shown in Fig. 1A. As intended, the controls scored nearly perfectly (mean = 90 ± 3% correct). In contrast, each patient answered only two questions correctly. The controls also performed well on the multiple-choice version of the test (97 ± 2% correct; Fig. 1B). In contrast, the patients each scored 55% correct, a score not reliably above the chance score of 33% (binomial test, P = 0.18). Both patients were questioned about their correct answers, but neither patient was able to provide correct information, and it seemed possible that their correct answers may have been "well-educated guesses". For example, on the recall test, both patients stated correctly that O.J. Simpson was charged with murder ("What crime was O.J. Simpson charged with?"). When E.P. was asked for more information, he stated, "No, I'm blank." G.P. stated that O.J. Simpson was associated with "murder, his wife," and then incorrectly added "I think he is in jail" and "I think she was shot. She was not strangled or stabbed." In addition, on the multiple-choice test, E.P. and G.P. correctly chose Afghanistan as the country that was invaded in an effort to capture Osama

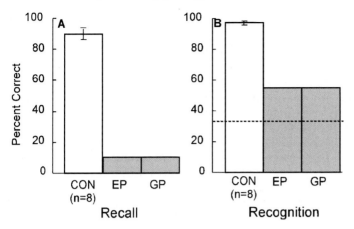

Fig. 1. Twenty easy facts. Patients E.P., G.P., and controls (CON) were given 20 questions about information that could only have been acquired after 1988. (**A**) Percent correct recall. (**B**) Percent correct responses on a three-alternative, forced-choice recognition test. Brackets show standard error of the mean, and the dashed line indicates chance performance (33%)

Bin Laden (alternatives, Algeria and Colombia). However, neither patient was able to provide any information about who this person was or why he had become well known.

Patients and controls were also given a news events test in which 55 questions were asked about notable events that had been in the news from 1990 to 2000 (e.g., "What is the name of the Russian leader who succeeded Boris Yeltsin?"). Testing was given first in a free-recall format and then in a four-alternative, multiple-choice format. The controls scored 51 ± 6% correct on the news events recall test (Fig. 2A). In contrast, patients E.P. and G.P. performed poorly (1.8% and 3.6% correct, respectively). Again, the patients were not able to provide any correct details about their correct answers and, in some instances, appeared to be relying on semantic knowledge gained early in life. Thus, when E.P. was asked which country invaded Kuwait, he replied "Iran and Iraq." When asked to choose, he said "Iraq" and expressed surprise when told that his answer was correct. The controls scored 85 ± 3% correct on the recognition test, whereas both patients performed near the chance score of 25% (E.P., 32% correct; G.P., 26% correct, P > 0.10; Fig. 2B). Again, the patients knew little, if anything, about their few correct answers.

The patients and controls were also given a test of famous faces, consisting of 24 photographs of famous people who came into the news in the 1990s. The test was given first in a free-recall format ("Who is this famous person?" e.g., Tiger Woods, Bob Dole). After the recall test, they were shown each photograph that they could not identify and asked either a yes-no question (e.g., "Is this person's name Tiger Woods?" The name that was presented was correct half the time) or a three-alternative, multiple-choice question (e.g., "Which of the following is the correct name?" Bob Dole, Newt Gingrich, Ross Perot). Controls scored 58 ± 8% correct on the famous faces recall test (Fig. 3A). In contrast, E.P. failed to recall a single name, and G.P. identified only one face correctly out of 24 (Colin Powell). On the recognition test (Fig. 3B), the controls scored 91 ± 2%

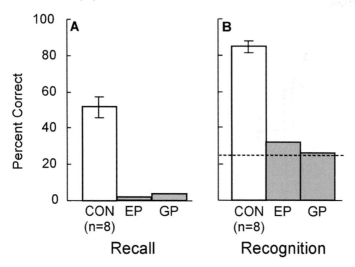

Fig. 2. News events. Patients E.P., G.P., and controls (CON) were given 55 questions about news events that occurred from 1990 to 2000. (**A**) Percent correct recall. (**B**) Percent correct responses on a four-alternative, forced-choice recognition test. Brackets show standard error of the mean, and the dashed line indicates chance performance (25%)

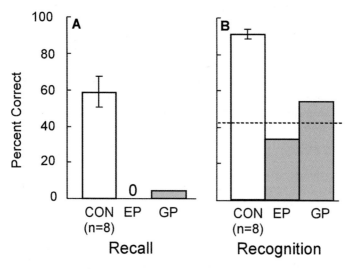

Fig. 3. Famous faces. Patients E.P., G.P., and controls (CON) were shown photographs of 24 persons who had become famous during the 1990s. (**A**) Percent correct recall of the name of the person. (**B**) Percent correct responses for recognition (both three-alternative, forced-choice recognition and yes/no recognition). Brackets show standard error of the mean, and the dashed line indicates chance performance (42%)

correct. In contrast, both patients performed near the chance score of 42% (E.P. 33% correct, P > 0.10; G.P., 54% correct, P > 0.10). Again, the patients could provide no accurate information about their few correct answers.

Participants were next given a test of famous persons who were either living or had died recently. For this test, participants were read aloud the names of famous (n = 126) and fictitious (n = 126) people, one at a time. All the famous people had become famous by 1970, and about one half (56%) of these individuals had died between 1990 and 2001. Participants first judged whether each name was the name of a famous person. Then, for each person correctly judged to be famous, participants were asked whether the person was still living or was deceased. To eliminate the use of age as a cue, the average age of the famous people who were living was the same as the age that the deceased people would have reached had they still been living. When judging whether a name was famous or not famous, controls scored 96 ± 1% correct, and patients E.P. and G.P. scored 73% and 78% correct, respectively (d' discriminability score = 3.8 ± 0.6 for controls; 1.2 for E.P. and 1.5 for G.P.). Figure 4 shows performance when participants were asked to judge whether the persons correctly judged to be famous were still living. Controls scored 73 ± 3% correct. In contrast, E.P. and G.P. scored near chance, 53% and 50% correct, respectively (chance = 50%). Thus the patients had virtually no knowledge of which famous persons were living and which famous persons had died during the past 10 years.

It could be argued that the failure of the patients to show new semantic learning is due to the fact that they have had limited exposure to sources of information about the world since they became amnesic and that the controls have had much more access to information during the corresponding time period. In the final test of new semantic learning, the patients were asked to draw a floor plan of their current residence from memory. Both patients had moved into their current residence (E.P. in 1993, G.P. in 2003) after the onset of their amnesia. Thus, each patient must have had thousands of opportunities to learn about the layout of their house, and any knowledge of the floor plan would represent new learning. Figure 5 shows the drawings made by E.P. and G.P. E.P.'s drawing correctly indicated the overall orientation of his house relative to the front door and showed correctly that the house was divided by a central corridor. However, the location of individual rooms was highly inaccurate. G.P.'s drawing appeared to be

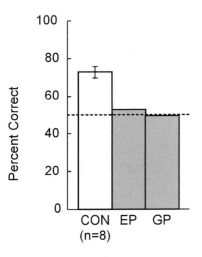

Fig. 4. Living or nonliving famous persons. Patients E.P., G.P., and controls (CON) were first presented with the names of 126 people who had become famous before 1970 and 126 not famous names and asked to judge which persons were famous. For each person correctly judged to be famous, they were then asked whether that person was still living (44% of the famous persons were still living, and 56% had died between 1990 and 2001). The bars show the percent correct score for the living/nonliving judgment. Brackets show standard error of the mean, and the dashed line indicates chance performance (50%)

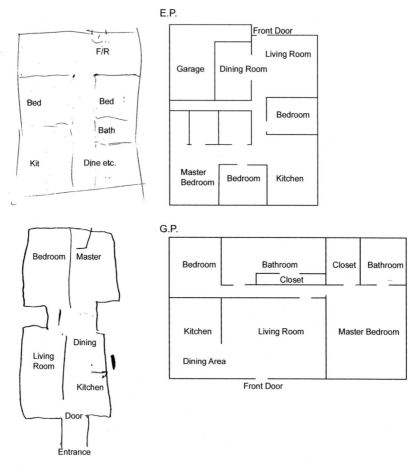

Fig. 5. Home floor plan. Patients E.P. (*top*) and G.P. (*bottom*) were asked to draw a floor plan of their current home from memory. Both patients had moved to their current homes after the onset of their amnesia. An accurate floor plan is shown alongside each patient's drawing

even less accurate. The overall shape of the residence was incorrect relative to the front door. The location of rooms was highly inaccurate.

Together, these findings indicate that patients E.P. and G.P. have acquired little, if any, new semantic knowledge about the world since the onset of their amnesia. Thus, the ability to acquire conscious knowledge across widely varying learning conditions appears to be dependent on the medial temporal lobe. This finding suggests that, when amnesic patients with extensive medial temporal lobe damage demonstrate some new semantic learning after the onset of their amnesia (Kitchener et al. 1998; O'Kane et al. 2004; Verfaellie et al. 2000, Westmacott and Moscovitch 2001), the new learning is supported by medial temporal lobe structures that remain intact.

The question arises as to why intact nondeclarative memory did not support the learning of some new semantic information. For example, in a previous study, E.P. was found to be capable of learning fact-like knowledge when given extensive training over

12 weeks (Bayley and Squire 2002). Similarly, other amnesic patients have demonstrated some capacity for new learning using a similar training schedule (Stark et al. 2005). The answer appears to be that nondeclarative memory can proceed under highly structured learning circumstances, in which stimuli are presented repeatedly in exactly the same way each time. However, nondeclarative memory is unsuited to support learning under the highly variable conditions that are encountered outside the laboratory.

Spared memory for remote autobiographical events following damage to the medial temporal lobe

The work summarized in the previous section demonstrated that patients with near complete damage to the medial temporal lobe are unable to acquire new semantic knowledge. Nevertheless, these patients have a rich store of semantic memories that were acquired from their early life, before they became amnesic. This section summarizes recent work that has assessed the status of remote autobiographical (episodic) memories of patients E.P. and G.P. and other patients with damage thought to be restricted to the hippocampal region (hippocampus proper, dentate gyrus, and subicular complex).

Earlier studies suggested that remote autobiographical memories were intact after damage to the hippocampal region and surrounding cortex. For example, patient R.B., who had histologically identified lesions restricted to the CA1 field of the hippocampus bilaterally, was as good as controls at recalling autobiographical memories from his earlier life (Zola-Morgan et al. 1986). Similarly, patients L.M. and W.H. were unimpaired when asked to recall remote autobiographical memories (MacKinnon and Squire 1989). These two patients had histologically identified damage to all the CA fields of the hippocampus, as well as damage to the dentate gyrus, the subiculum (W.H. only), and some cell loss in the entorhinal cortex (Rempel-Clower et al. 1996). In these studies, the autobiographical memories were assessed using a 0–3 scoring system. In contrast to these findings, other patients have been described who appear to have difficulty recalling autobiographical episodes from all periods of their life, including the most remote time periods tested (Hirano and Noguchi 1998; Moscovitch et al. 2000; Cipolotti et al. 2001). Such findings raise the possibility that autobiographical memory is always dependent on the medial temporal lobe and that medial temporal lobe structures are always necessary for recollecting the richness of detail that is the hallmark of autobiographical memory.

There appear to be two ways to reconcile the available data. One possibility is that an impairment of remote autobiographical memory was not detected in our earlier studies because we used only a simple 0–3 scoring system. A patient's narrative could receive a full score of 3, yet still be impoverished in detail. A second possibility is that those patients who are reported to perform poorly on tests of remote autobiographical memory might have significant damage to critical brain structures outside the medial temporal lobe. We assessed these possibilities in two studies. In the first study (Bayley et al. 2003), we carried out a detailed analysis of the narrative content of remote autobiographical memories elicited from six well-characterized amnesic patients who have damage limited to the hippocampal region and two patients (E.P. and G.P.) who (as described earlier) have extensive damage to the medial temporal lobe. In a second study (Bayley et al. 2005b), we carried out a detailed volumetric analysis of structural brain images from eight memory-impaired patients. Five of the patients have damage limited to the medial temporal lobe, and three have medial temporal lobe damage plus significant additional damage to neocortex. The aim of this second study was to identify anatomical differences between patients who could successfully recall remote autobiographical memories and those who could not.

In the first study (Bayley et al. 2003), eight patients were each asked to recollect autobiographical memories from the first third of their lives before the onset of their amnesia (on average, before the age of 16 years; Fig. 6). Twenty-five age-matched controls were also asked to recollect memories from the corresponding portion of their early lives. Each participant was given 24 cue words, one at a time (e.g., river, bird, nail), and asked to produce a memory involving the cue word that was specific in time and place (Crovitz and Schiffman 1974). Participants were prompted to elicit as much detail as possible. The responses were tape-recorded, and those responses that involved a specific event were later scored according to how many details were contained in each narrative. More than 700 narratives were evaluated. Both the patients and controls were able to provide specific recollections in response to the cue words (patients, 89.2% of the words; controls 95.3%; $P > 0.10$). Two categories of detail were defined: episodic detail and semantic detail. An episodic detail was specific to an event (e.g., the outcome of a particular sporting event played at school one day). In contrast, a semantic detail was not unique to the event being recollected but was nonetheless

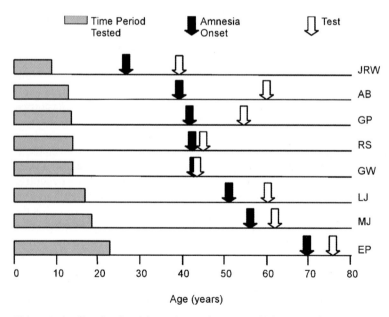

Fig. 6. A timeline for the eight patients who were asked to provide remote autobiographical memories. The figure illustrates the time period from which each patient was asked to draw remote autobiographical memories, the patient's age at the onset of amnesia, and age at the time of testing

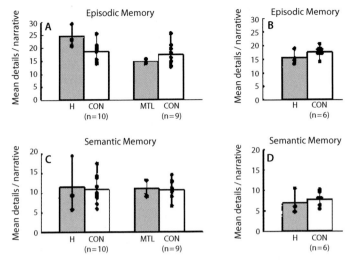

Fig. 7. The number of details contained in remote autobiographical memories. Each participant was given 24 cue words (e.g., river, bottle, nail) one at a time, and asked to recollect a specific event from the first third of life that involved the word. Narratives were collected in two separate studies by different interviewers. *Panels* (**A**) (first study) and (**B**) (second study) show the mean number of details per narrative that described specific autobiographical events (episodic memory). *Panels* (**C**) (first study) and (**D**) (second study) show the mean number of details per narrative that were recounted as part of the recollections but were not unique to a specific episode (semantic memory). Each participant is represented by a filled circle, and patients are identified by their initials. H = patients with lesions thought to be limited to the hippocampal region (hippocampus proper, dentate gyrus, and subicular complex); MTL = post-encephalitic patients with large medial temporal lobe lesions; CON = controls

relevant to the narrative (e.g., the name of the school). Figure 7 illustrates the results, shown as two separate studies. (The two studies were identical except that they were conducted by different interviewers.) The interviewer for the second study did not probe for details as extensively as the interviewer for the first study, which resulted in somewhat fewer elicited details for controls and patients alike in the second study). The recollections of the patients and the controls contained a similar number of details (±5%). Approximately two-thirds of the total details were classified as episodic details, and the remaining one-third were classified as semantic details. The patients and the controls also performed similarly in a number of other respects. They took a similar amount of time to begin their narratives, received a similar number of prompts before beginning, and took a similar amount of time to report the narratives.

The reliability and accuracy of the methods used in the study were also checked. To assess reliability, the narratives from 10 of the participants (four patients, six controls) were scored independently by two raters. Both raters assigned the amnesic patients and controls a similar number of details (for both raters, < 1.0 detail/narrative separated the patients and the controls). An effort was also made to determine the accuracy of the narratives. After the initial interview (median = 14 months), the patients and controls were provided a few cues for each of their recollections and were then asked four specific questions about each narrative that was specific to a particular time and place (maximum = 24). The rationale was that participants (and especially the patients) would have difficulty answering questions about a narrative if it had been manufactured at the time of the original interview. Using this method, 88.0% of the patient recollections and 93.4% of the control recollections were considered to be confirmed ($P > 0.10$).

These results indicate that patients with lesions to the medial temporal lobe are able to recollect detailed, remote autobiographical memories as well as controls do. This conclusion should be qualified in two ways. First, it is not possible to conclude that the autobiographical memories of the patients are intact in every respect. Although we were unable to find a difference, either qualitatively or quantitatively, between the patient and control memories, it remains possible that the two groups differed in some way that was not detected by our analysis. However, the results clearly rule out the possibility that patients with medial temporal lobe lesions are grossly deficient at recollecting remote autobiographical events.

Second, it is difficult to determine to what extent the recollections provided by the patients can be called episodic memories in the same sense that healthy individuals can recall episodes from their recent past. However, it is also difficult to make this determination for recollections provided by the control group. One interesting possibility is that autobiographical memories become more fact-like and less episodic as they become more remote. For example, Cermak (1984) proposed that remote memory in amnesia consists mostly of frequently told stories that have become disconnected from their original context and are now part of semantic knowledge. If this proposal applies to patients with damage restricted to the medial temporal lobe, we would suggest that the same notion also applies to healthy individuals.

Although the study described above found that the memories provided by the patients were as rich and detailed as the memories provided by the controls, others have suggested that autobiographical memory remains dependent on the medial temporal lobe for as long as the memory persists (Fujii et al. 2000; Rosenbaum et al. 2001). Accordingly, in a second study (Bayley et al. 2005b), we investigated the locus and extent of damage in the patients who can successfully recollect autobiographical memories and compared this damage to that in patients who cannot recollect autobiographical memories. In this study, a detailed volumetric analysis was performed on the magnetic resonance images from memory-impaired patients and controls. Five of the patients had taken part in the study just described (Bayley et al. 2003) and had damage limited to the medial temporal lobe (designated as the "MTL" group). These five patients could recollect remote autobiographical memories as successfully as their controls. Three other patients also took part. They were known to have medial temporal lobe damage plus significant additional damage to neocortex (the "MTL+" group). These three patients were strikingly impaired in the recollection of remote autobiographical memory.

Autobiographical memories were elicited using the methods described above (Bayley et al. 2003), in which a list of 24 high-frequency nouns were presented one at a time with the instruction to recollect a unique event that involved the stimulus word and that was specific to time and place. Narratives were tape-recorded and later scored on a 0 to 3 scale. Three points were awarded for an episodic memory that was specific to time and place (e.g., a description of the day the participant started at a new school).

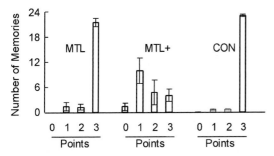

Fig. 8. Performance on a test of remote autobiographical memory. Participants were given 24 cue words (e.g., river, bottle, nail) and asked to recollect a specific event that was associated with each word. Patients were asked to recall events from the first third of their life before the onset of amnesia. Controls were asked for events from corresponding time periods. Narratives were scored (0 to 3) for how well they described an event (0 = no response or a generic response, 1 = vague reference to a memory without any reference to time or place, 2 = memory had some specificity but was not specific to one time and place, 3 = memory was specific to one time and place). The bars show the mean number of narratives given each score. MTL, six patients with hippocampal lesions or larger medial temporal lobe lesions; MTL+, three patients with medial temporal lobe lesions and significant additional damage to neocortex; CON, 26 controls

Two points were awarded for a memory that had some specificity but was not specific in time and place and was therefore not recalled as a unique event (e.g., "I used to stay at my grandma's house on weekends"). One point was awarded for a vague reference (e.g., "I read a lot of books"). Zero points were given for no response or a generic response (e.g., "You can open and close a door").

The quality of the autobiographical memories of the two patient groups (MTL and MTL+) and 26 controls are illustrated in Fig. 8. The patients in the MTL group and controls were able to provide unique autobiographical memories (score of 3 points) in response to most of the 24 cue words (MTL patients, 21.6 memories; controls, 22.9 memories). In contrast, the three patients in the MTL+ group provided a mean of only 4.0 unique autobiographical memories in response to the 24 cue words. The majority of the memories provided by the MTL+ group were awarded just one point, suggesting that these patients were able to recall general information but had difficulty providing memories that were specific to time and place.

For the volumetric analysis of brain images, two medial temporal lobe structures were defined for each hemisphere: the hippocampal region (hippocampus proper, dentate gyrus, and subicular complex) and the parahippocampal gyrus (perirhinal, entorhinal, and parahippocampal cortices). To obtain volumetric data for the neocortex, twelve regions of interest were defined, including the frontal lobes, lateral temporal lobes, parietal lobes, occipital lobes, insular cortex and fusiform gyrus (both left and right sides). Twelve controls were matched to the patients and also analyzed.

In the MTL group, three of the patients (R.S., G.W., J.R.W.) were found to have substantial volume reductions within the hippocampal region but, with one exception, no reduction in the parahippocampal gyrus, the fusiform gyrus, the insular cortex, or the major lobes of the brain. The one exception was patient R.S., whose parietal lobes are unusually small. This finding likely reflects natural variation in the parietal lobe volume rather than damage to this brain region for the following reasons: 1) the parietal lobes are highly variable in size (Raz et al. 2005); 2) no evidence of parietal lobe damage is apparent in R.S.'s MRI scan; and 3) R.S. obtained normal scores on tests sensitive to parietal lobe function. Patients E.P. and G.P. have more extensive medial temporal lobe damage than the other patients in the MTL group. Specifically, the hippocampal region and parahippocampal gyrus are markedly reduced in volume bilaterally. The major lobes of the brain were of normal volume.

In contrast to the patients in the MTL group, patients in the MTL+ group were found to have reduced volumes of medial temporal lobe structures and additional, significant reductions in the volume of one or more of the major lobes. Specifically, H.C. has reduced volumes of the frontal, parietal and occipital lobes. These reduced volumes are bilateral, although the volume reduction in H.C.'s frontal lobe reached significance only in the right hemisphere. Patient P.H. was found to have reduced volume in the left frontal lobe and a similar, albeit not significant, reduction in the right frontal lobe. Lastly, patient G.T. has reduced volumes in the lateral temporal lobes bilaterally.

The major finding was that patients with significant damage to neocortex were markedly impaired on tests of remote autobiographical memory, whereas patients with damage restricted to the medial temporal lobe performed normally. The three patients who had difficulty in recalling remote memories all had damage to areas of the cortex that have been implicated in remote memory. For example, patient G.T. had damage to the lateral temporal cortex, which when damaged is known to impair remote autobiographical memory (Graham and Hodges 1997). Further, patients H.C. and P.H. both had frontal lobe damage. The frontal lobe is known to be important for a variety of "executive" functions that are important for the strategic aspects of recall and for active or effortful reconstructive processes (Kopelman 2002). Furthermore, frontal lobe damage has been reported to impair autobiographical memory (Kopelman et al. 2003). Lastly, patient H.C. had damage to the occipital lobe. The occipital region has been associated with the retrieval of visual images, and damage to this region has also been associated with impaired autobiographical recollection (Rubin and Greenberg 1998).

As described above, some memory-impaired patients have been reported to do poorly at recollecting autobiographical memory, and these impairments have been attributed to their medial temporal lobe damage (patient V.C., Cipolotti et al. 2001; patient H.M., Steinvorth et al. 2005; patient Y.K., Hirano et al. 2002). On the basis of the anatomical evidence considered above, it is important to also consider the possibility that these patients have significant damage outside the medial temporal lobe. Another possibility is that the differences reported in different studies are due to important variations in test procedures. Indeed, it has been proposed that, until standardized techniques are used across laboratories, differences in how remote memories are assessed could explain differences in findings (Rosenbaum et al. 2004). However, it should be noted that a simple, standardized procedure has been used to assess the autobiographical memories of many of the patients under study [the Autobiographical Memory Interview (AMI); Kopelman et al. 1989]. The AMI assesses autobiographical memory for childhood events using three questions (maximum score = 9 points).

The crucial finding is that patients who reportedly do poorly at producing detailed remote autobiographical memories also perform poorly on the childhood portion of the AMI [patient Y.K. = 4/9 points (Hirano and Noguchi 1998); patient V.C. = 1/9 points (Cipolotti et al. 2001)]. Notably, our patients in the MTL+ group, who have damage outside the medial temporal lobe, all failed to reach a maximum score

[H.C. = 7/9 points; G.T. = 0/9 points; P.H. = 0/9 points (see Bayley et al. 2005b)]. In contrast, our five patients in the MLT group who had damage limited mainly to the medial temporal lobe all achieved the maximum score of nine points on the childhood portion of the AMI. These results indicate that, even when a simple standardized test is used, one still finds differences among patients in their ability to recollect remote autobiographical memories. Accordingly, the different results recorded in the literature cannot be a function of differences in the test procedures that have been used.

We suggest that whenever memory-impaired patients have significant difficulty in recollecting remote autobiographical memories, they will ultimately prove to have damage outside the medial temporal lobe. The ability to retrieve remote autobiographical memories is supported by neocortex, including the frontal, lateral temporal and occipital lobes. It should be noted that these conclusions are consistent with findings from many studies using experimental animals. In these studies, after lesions to the hippocampus or related structures, information learned long before the onset of memory impairment is spared relative to information learned recently (Squire et al. 2004). Animal studies have also demonstrated the increasing importance of the neocortex as memories grow older (Frankland et al. 2004; Maviel et al. 2004; Wiltgen et al. 2004; Frankland and Bontempi 2005). The present study demonstrates that human memory follows a similar pattern: remote autobiographical memory is supported by the neocortex and is independent of the medial temporal lobe.

Robust Habit Learning in Humans

Habit memory is a form of nondeclarative memory that involves slowly acquired associations between stimuli and responses. In contrast to declarative memory, nondeclarative memory operates outside of awareness and is expressed through performance rather than conscious recollection. Declarative memory depends on the integrity of the medial temporal lobe. By contrast, habit memory is thought to be supported by the basal ganglia (Mishkin and Petri 1984). It has been difficult to study habit learning in humans because humans tend to spontaneously engage their declarative memory when performing memory tasks. In contrast, experimental animals readily and spontaneously learn by habit, even when given a memory task that, in humans, engages declarative learning.

The issue is nicely illustrated by the concurrent discrimination task (Hayes et al. 1953; Correll and Scoville 1965). In this task, eight pairs of objects are typically presented five times per day, one pair at a time. One of the objects in each pair is arbitrarily designated as always correct, and the choice of that object produces a reward. Humans quickly learn this task using declarative memory and perform at approximately 90% correct after one or two training days. Amnesic patients are markedly impaired at learning the task (Squire et al. 1988; Hood et al. 1999). The findings are different in the monkey. First, normal monkeys learn the concurrent discrimination task only gradually and require several hundred trials to learn all eight pairs. Further, monkeys with either hippocampal lesions or large medial temporal lobe lesions acquire the task at a normal rate (Buffalo et al. 1998; Malamut et al. 1984; Teng et al. 2000). Lastly, in the monkey, damage to the caudate nucleus impairs performance on this task (Teng et al. 2000; Fernandez-Ruiz et al. 2001).

One possibility is that habit learning is relatively underdeveloped in humans such that habit learning cannot support the learning of concurrent discrimination, even though the task is readily learned by animals as a habit. Alternatively, habit memory may be capable of supporting this kind of learning in humans, but habit learning may be overridden by a tendency to engage a declarative memory strategy. We explored these possibilities by giving the concurrent discrimination test to two profoundly amnesic patients, E.P. and G.P., and four matched controls (Bayley et al. 2005a). As described earlier, these two patients are virtually unable to acquire any new declarative knowledge (Stefanacci et al. 2000; Levy et al. 2004; Bayley and Squire 2005).

The concurrent discrimination test consisted of eight pairs of junk objects (miscellaneous pieces of metal or plastic). Testing was given twice each week (40 trials per session), and participants were required to learn by trial and error which object in each pair was correct. Specifically, they were required to pick up one of the two objects and discover whether the word "correct" appeared under its cardboard base. Figure 9A shows that the controls learned the task quickly and, at the end of three test days, scored 95.3% correct on average. The patients were markedly impaired but eventually succeeded at learning. E.P. learned gradually at a nearly linear rate (linear trend: $F_{(1,35)}$ = 54.7, P < 0.001). After 36 test sessions (18 weeks), he performed at 85.0% correct (Fig. 9B). G.P. showed a similar pattern of gradual learning (linear trend: $F_{(1,27)}$ = 13.6, P < 0.001) and after 28 test sessions (14 weeks) achieved a score of 92.5% correct (Fig. 9C).

Before each session, the patients were asked to describe what they were about to do. Strikingly, neither patient could report any specific knowledge about the test. For example, E.P. was never able to say anything more specific than that he and the experimenter would have a conversation. And when told that the test involved objects, he typically suggested that he and the experimenter would discuss their use. Yet, in the later sessions, E.P. would conclude his comments, turn to the test, and obtain high scores. It is also notable that the comments offered by the patients during testing described an automatic kind of responding. For instance, right at the end of session 34, E.P. was asked to describe his strategy for choosing the correct object. (He had just performed 80% correct). He stated, "It's just up here [pointing to head] from the memory. It seems like it's up there, and comes down and out".

One hallmark of habit memory is that it is inflexible and is thought to be dependent on the original training conditions for full expression (Mishkin and Petri 1984; Wise 1996). To test whether the knowledge acquired by the patients was inflexible, we administered a sorting task three to six days after the conclusion of formal training. For this task, all 16 objects were placed together on a table. Participants were told that some of the objects had been consistently designated as correct, and they were asked to sort the objects, placing the correct objects to one side and the other objects to the other side. Figure 9A shows that the controls performed well on the sorting task, scoring 95.3% correct. In contrast, E.P. and G.P. both failed the sorting task (56.3% and 50.0% correct, respectively, chance performance = 50%; Fig. 9B,C). Immediately afterwards, the original task was given to participants with instructions to verbalize their choices rather than to reach for the objects. All participants scored 90% or better. Seventeen days later, each patient was given the sorting task again, followed by 40 trials of the standard task (Fig. 9B,C). The results were the same as before. E.P. and G.P. failed

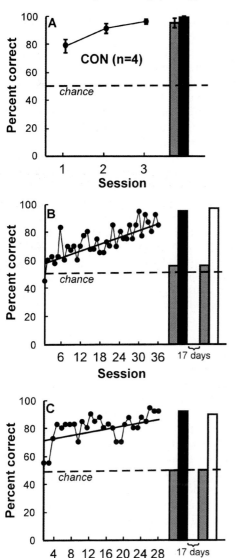

Fig. 9. Performance on the concurrent discrimination task. (**A**) Controls (CON) rapidly learned the task over three sessions and performed well on the sorting task three to six days later (*gray bar*). The black bar shows their performance immediately afterwards when they were required to verbalize their choices rather than reach for objects. Brackets show S.E.M. (**B**) E.P. gradually learned the task across 18 weeks. Five days later, he failed the sorting task (gray bar) but immediately afterwards, performed well in the standard task format while verbalizing his responses (black bar). Seventeen days later, E.P. again failed the sorting task (gray bar) but performed well when the original conditions were re-established. (**C**) G. P. learned the task gradually during 14 weeks. He failed the sorting task on two different occasions, five days after training and again 17 days later. Like E.P., he performed well immediately afterwards when the original task format was reinstated (*black bar*: verbalizing; *white bar*: standard task)

the sorting task (56.3% and 50.0% correct, respectively) but immediately afterwards succeeded at the standard task (97.5% and 90.0% correct, respectively).

Taken together, this study illustrates that habit learning in humans can support tasks such as concurrent discrimination, which humans ordinarily learn as declarative memory. Indeed, the patients appear to learn the task rather in the way that monkeys learn. They improve gradually over many hundreds of trials. However, learning under these circumstances operates outside of awareness and independently of the medial temporal lobe. What is learned is rigidly organized, and the expression of what has been learned depends on reinstating the original task format.

Conclusion

The studies reviewed in this chapter provide evidence for the distinction between declarative (conscious) and nondeclarative (unconscious) learning systems. In particular, the results indicate that the medial temporal lobe is essential for the acquisition of new declarative memory. Thus, patients with nearly complete damage to the medial temporal lobe fail to acquire new declarative knowledge. Nevertheless, these same patients, and other patients who have damage limited to the hippocampal region, are able to recollect detailed autobiographical memories from their remote past. Study of other patients who fail tests of remote autobiographical memory strongly suggests that widespread areas of the neocortex are the permanent repositories of autobiographical memories. In contrast to these observations of declarative memory, nondeclarative memory operates unconsciously and independently of the medial temporal lobe. Even patients with virtually no capacity to acquire new declarative memories can nonetheless acquire new information through habit learning. In sharp contrast to declarative memory, habit learning is slow, unconscious and rigidly organized.

References

Bayley PJ, Squire LR (2002) Medial temporal lobe amnesia: gradual acquisition of factual information by nondeclarative memory. J Neurosci 22:5741–5748

Bayley PJ, Squire LR (2005) Failure to acquire new semantic knowledge in patients with large medial temporal lobe lesions. Hippocampus 15:273–280

Bayley PJ, Hopkins RO, Squire LR (2003) Successful recollection of remote autobiographical memories by amnesic patients with medial temporal lobe lesions. Neuron 38:135–144

Bayley PJ, Frascino JC, Squire LR (2005a) Robust habit learning in the absence of awareness and independent of the medial temporal lobe. Nature 436:550–553

Bayley PJ, Gold JJ, Hopkins RO, Squire LR (2005b) The neuroanatomy of remote memory. Neuron 46:799–810

Buffalo EA, Stefanacci L, Squire LR, Zola SM (1998) A reexamination of the concurrent discrimination learning task: the importance of anterior inferotemporal cortex, area TE. Behav. Neurosci 112:3–14

Cermak LS (1984) The episodic-semantic distinction in amnesia. In: Squire LR, Butters N (eds) Neuropsychology of memory. Guilford Press, New York, pp 55–62

Cipolotti L, Shallice T, Chan D, Fox N, Scahill R, Harrison G, Stevens J, Rudge P (2001) Long-term retrograde amnesia: the crucial role of the hippocampus. Neuropsychologia 39, 151–172

Correll RE, Scoville WB (1965) Performance on delayed match following lesions of medial temporal lobe structures. J Comp Physiol Psych 60:360–367

Crovitz HF, Schiffman H (1974) Frequency of episodic memories as a function of their age. Bull Psychon Soc 4:517–518

Eichenbaum H, Cohen NJ (2001) From conditioning to conscious recollection: memory systems of the brain. Oxford University Press, New York

Fernandez-Ruiz J, WangJ, Thomas G. Aigner TG, Mishkin M (2001) Intact visual perception in memory-impaired patients with medial temporal lobe lesions. Proc Natl Acad Sci USA 98:4196–4201

Frankland PW, Bontempi B (2005) The organization of recent and remote memories. Nature Rev Neurosci 6:119–130

Frankland PW, Bontempi B, Talton LE, Kaczmarek L, Silva AJ (2004) The involvement of the anterior cingulate cortex in remote contextual fear memory. Science 304:881–883

Fujii T, Moscovitch M, Nadel L (2000) Memory consolidation, retrograde amnesia, and the temporal lobe. In: Cermak L (ed) Handbook of neuropsychology. Elsevier, Amsterdam, pp 223–250

Graham KS, Hodges JR (1997) Differentiating the roles of the hippocampal complex and the neocortex in long-term memory storage: evidence from the study of semantic dementia and Alzheimer's disease. Neuropsychology 11:77–89

Hayes KJ, Thompson R, Hayes C (1953) Concurrent discrimination learning in chimpanzees. J Comp Physiol Psych 46:105–107

Hirano M, Noguchi K (1998) Dissociation between specific personal episodes and other aspects of remote memory in a patient with hippocampal amnesia. Percept Mot Skills 87:99–107

Hirano M, Noguchi K, Hosokawa T, Takayama T (2002) I cannot remember, but I know my past events: remembering and knowing in a patient with anmesic syndrome. J Clin Exp Neuropsych 24:548–555

Hood KL, Postle BR, Corkin S (1999) An evaluation of the concurrent discrimination task as a measure of habit learning: performance of amnesic subjects. Neuropsychologia 37:1375–1386

Kitchener EG, Hodges JR, McCarthy R (1998) Acquisition of post-morbid vocabulary and semantic facts in the absence of episodic memory. Brain 121: 1313–1327

Kopelman MD (2002) Disorders of memory. Brain 125:2152–2190

Kopelman MD, Wilson BA, Baddeley AD (1989) The autobiographical memory interview: a new assessment of autobiographical and personal semantic memory in amnesic patients. J Clin Exp Neuropsych 5:724–744

Kopelman MD, Lasserson D, Kingsley DR, Bello F, Rush C, Stanhope N, Stevens TG, Goodman G, Buckman JR, Heilpern G, Kendall BE, Colchester AC (2003) Retrograde amnesia and the volume of critical brain structures. Hippocampus 13:879–891

Levy D, Bayley PJ, Squire LR (2004) The anatomy of semantic knowledge: medial vs. lateral temporal lobe. Proc Natl Acad Sci USA 101:6710–6715

MacKinnon D, Squire LR (1989) Autobiographical memory in amnesia. Psychobiology 17:247–256

Malamut BL, Saunders RC, Mishkin M (1984) Monkeys with combined amygdalo-hippocampal lesions succeed in object discrimination learning despite 24-hour intertrial intervals. Behav Neurosci 98:759–769

Maviel T, Durkin TP, Menzaghi F, Bontempi B (2004) Sites of neocortical reorganization critical for remote spatial memory. Science 305:96–99

McClelland JL, McNaughton BL, O'Reilly RC (1995) Why there are complimentary learning systems in the hippocampus and neocortex: insights from the success and failures of connectionist models of learning and memory. Psychol Rev 3:419–457

Mishkin M, Petri HL (1984) Memories and habits: some implications for the analysis of learning and retention In: Squire LR, Butters N (eds) Neuropsychology of memory. The Guilford Press, New York, pp 287–296

Moscovitch M, Yaschyshyn T, Ziegler M, Nadel L (2000) Remote episodic memory and retrograde amnesia: was Endel Tulving right all along? In: Tulving E (ed) Memory, consciousness and the brain: the Tallinn Conference. Psychology Press/Taylor & Francis, Philadelphia, pp 331–345

O'Kane G, Kensinger EH, Corkin S (2004) Evidence for semantic learning in profound amnesia: an investigation with the patient H.M. Hippocampus 14:417–425

Raz N, Lindenberger U, Rodrigue KM, Kennedy KM, Head D, Williamson A, Dahle C, Gerstorf D, AckerJD (2005) Regional brain changes in aging healthy adults: general trends, individual differences and modifiers. Cereb Cortex 15:1676–1689

Rempel-Clower N, Zola SM, Squire LR, Amaral DG (1996) Three cases of enduring memory impairment following bilateral damage limited to the hippocampal formation. J Neurosci 16:5233–5255

Rosenbaum RS, Winocur G, Moscovitch M (2001) New views on old memories: re-evaluating the role of the hippocampal complex. Behav Brain Res 127:183–197

Rosenbaum RS, McKinnon MC, Levine B, Moscovitch M (2004) Visual imagery deficits, impaired strategic retrieval, or memory loss: disentangling the nature of an amnesic person's autobiographical memory deficit. Neuropsychologia 42:1619–1635

Rubin DC, Greenberg DL (1998) Visual memory-deficit amnesia: a distinct amnesic presentation and etiology. Proc Natl Acad Sci USA 95:5413–5416

Shrager Y, Gold JJ, Hopkins R, Squire LR (2006) Intact visual perception in memory-impaired patients with medial temporal lobe lesions. J Neurosci 26:2235–2240

Squire LR, Zola-Morgan (1991) The medial temporal lobe memory system. Science 253:1380–1386

Squire LR, Zola-Morgan S, Chen K (1988) Human amnesia and animal models of amnesia: performance of amnesic patients on tests designed for the monkey. Behav Neurosci 11:210–221

Squire LR, Clark RE, Bayley PJ (2004) Medial temporal lobe function and memory. In: Gazzaniga (ed) The cognitive neurosciences III. The MIT Press, Cambridge Mass, pp 691–708

Stark CEL, Squire LR (2000) Intact visual perceptual discrimination in humans in the absence of perirhinal cortex. Learn Mem 7:273–278

Stark CEL, Stark S, Gordon B (2005) New semantic learning and generalization in an amnesic patient. Neuropsychology 19:139–151

Stefanacci L, Buffalo EA, Schmolck H, Squire LR (2000) Profound amnesia following damage to the medial temporal lobe: a neuroanatomical and neuopsychological profile of patient E.P. J Neurosci 20:7024–7036

Steinvorth S, Levine B, Corkin S (2005) Medial temporal lobe structures are needed to re-experience remote autobiographical memories: evidence from H.M. and W.R. Neuropsychologia 43:479–496

Teng E, Stefanacci L, Squire LR, Zola SM (2000) Contrasting effects on discrimination learning following hippocampal lesions or conjoint hippocampal-caudate lesions in monkeys. J Neurosci 20:3853–3863

Tulving E (1991) Concepts in human memory. In: Squire LR, Weinberger NM, Lynch G, McGaugh JL (eds) Memory: organization and locus of change. Oxford University Press, New York, pp 3–32

Verfaellie M, Koseff P, Alexander MP (2000) Acquisition of novel semantic information in amnesia: effects of lesion location. Neuropsychologia 38:484–492

Westmacott R, Moscovitch M (2001) Names and words without meaning: incidental postmorbid semantic learning in a person with extensive bilateral medial temporal lobe damage. Neuropsychology 15:586–596

Wiltgen BJ, Brown RA, Talton LE, Silva AJ (2004) New circuits for old memories: the role of the neocortex in consolidation. Neuron 44:101–108

Wise SP (1996) The role of the basal ganglia in procedural memory. Sem Neurosci 8:39–46

Zola-Morgan S, Squire LR, Amaral DG (1986) Human amnesia and the medial temporal region: enduring memory impairment following a bilateral lesion limited to field CA1 of the hippocampus. J Neurosci 6:2950–2967

Dynamics of Hippocampal-Cortical Interactions During Memory Consolidation: Insights from Functional Brain Imaging

Bruno Bontempi[1] and *Thomas P. Durkin*[1]

Summary

Both clinical studies and experiments in animals have provided evidence for the existence of a temporally graded retrograde amnesia following lesions of the medial temporal lobe, including the hippocampus. This form of amnesia, which is characterized by a loss of memory for recent events acquired prior to the onset of amnesia while more remote memories are preserved, is one of the major arguments for the existence of a consolidation process necessary for stable, long-term memory formation. It is now well established that the formation of declarative memory (memories for facts and events) involves changes in synaptic plasticity within the medial temporal lobe. However, our group and others have demonstrated that the hippocampus has only a time-limited role in long-term memory storage of certain types of information, such that extrahippocampal structures, namely cortical regions, eventually become capable of supporting the retrieval of remote memories independently. In other words, the hippocampus does not store remote memories, yet what happens beyond the hippocampus remains unclear. This issue has been the subject of intense investigation and debate in the field of cognitive neuroscience, but to date, no convincing evidence as to the identity, mechanisms and putative interactions between memory systems underlying remote memory storage and retrieval have clearly emerged. To address this issue, we have conducted brain imaging experiments using (^{14}C)2-deoxyglucose mapping and analyses of changes in the expression of activity-dependent genes (*c-fos* and *Zif268*) in mice submitted to recent and remote spatial memory testing. Our findings show that memory processing and consolidation require a time-dependent hippocampal-cortical dialogue, ultimately enabling structured cortical networks to mediate recall and use of cortically stored remote memories independently. However, the cortex does not simply serve as a passive storage site but may also actively integrate new memories depending on the organization and status of pre-existing knowledge. The prefrontal cortex in particular appears to play a crucial role in integrating and binding information from distributed cortical networks and in modulating the level of hippocampal activation during memory recall. These findings are discussed in the context of current models of memory consolidation and in light of data from the recent literature in humans and animals.

---

[1] Laboratoire de Neurosciences Cognitives, CNRS UMR 5106, Université de Bordeaux 1, Avenue des Facultés, 33405 Talence, France
b.bontempi@lnc.u-bordeaux1.fr

Bontempi et al.
Memories: Molecules and Circuits
© Springer-Verlag Berlin Heidelberg 2007

Introduction

One of the most astounding features of human memory is the fact that its capacity for persistence encompasses a time-scale ranging from milliseconds to the many decades that cover the life span of an individual (Squire and Kandel 2000). The idea that memory traces are not acquired in their definitive state at the moment of initial acquisition but rather undergo a gradual process of stabilization and consolidation over time has been a central concept in cognitive neuroscience for several decades and finds its roots in the clinical observations made by the French psychologist Theodule Ribot as early as 1881. Ribot was the first to relate memory loss after a brain insult to the age of the memory and to point out that recently acquired information was typically more impaired than remote memories. This dissociation led him to suggest the probable existence of a time-dependent process of memory reorganization, which later became known as Ribot's gradient. Building on major neuroanatomical and neuropsychological advances during the twentieth century, further investigations of memory function in brain-damaged patients and animals have provided remarkable insights into the brain regions underlying memory processes and into how recent and remote memories are organized and stored in the brain (McGaugh 2000; Dudai 2004; Squire et al. 2004; Frankland and Bontempi 2005; Moscovitch et al. 2006). In humans, the medial temporal lobe, which includes the hippocampus, has been shown to play a crucial role in processing memory for facts and events (i.e., "declarative memories"; Eichenbaum 2000, 2004). As these memories mature over time, they are thought to become progressively independent of the hippocampus and increasingly dependent on other brain regions such as the cortex (Squire and Alvarez 1995; McClelland et al. 1995; Squire et al. 2004). However, the contribution of these other brain regions and the neuronal mechanisms underlying the consolidation process have remained elusive, and this issue constitutes a major research topic in the field. Recent studies have begun to shed light on the dynamics of hippocampal-cortical interactions that enables recently acquired memories to be transformed into enduring memories in distributed cortical networks. In this chapter, we examine these recent advances, starting with a brief history of research into memory consolidation that has led to the proposal of two main, but divergent, theories concerning the role of the hippocampus in memory consolidation. Experimental evidence in support of each of these theories is reviewed in light of the existing literature and also addressed by recent functional brain imaging studies conducted by our group using mice. The respective roles of hippocampal and cortical brain regions in processing recently versus remotely acquired information and the mechanisms involved in cortical storage are also briefly discussed, with a particular focus on the role of the prefrontal cortex, which we suggest progressively exerts a privileged role during the storage, organization and retrieval of remote memories.

Evolution of the concept of memory consolidation

A brief history

The first authors to have used the term "consolidation" were Müller and Pilzecker (1900). They formulated the hypothesis that information processing within the brain remains sensitive to interference during a certain time-period up until the point when the memory traces become "consolidated", and thereby resistant to disruption. Importantly, these same authors also suggested that long-term memory storage of information (in their study, verbal material) would be mediated via the temporary persistence of "certain physiological processes" – nowadays commonly referred to as patterns of neuronal activation – induced by the acquisition of that information (see also Lechner et al. 1999). In 1903, Burnham, basing his observations on cases of retrograde amnesia, that is, a loss of information that was acquired prior to the onset of the amnestic event, described memory formation and consolidation as requiring a time-dependent process of repetition and association, and also included the notion of a reorganization and modification of the neuronal substrates underlying the storage of recently acquired information in order for by to integrate into long-term memory. The confirmation by Woodworth (1929) of retroactive interference (disruption of retention of a learning episode A by the later acquisition of another similar learning episode B, which Müller and Pilzecker initially called "retroactive inhibition") also provided additional support for the existence of the consolidation phenomenon and suggested that the neuronal processes engaged in the brain by a learning episode persist long after the end of the learning session. Karl Lashley was another pioneer in the search for the memory trace classically referred to as the "engram". By performing cortical lesions of different sizes in rats confronting cognitive challenges in complex mazes, Lashley (1950) concluded that memories are not localized but are rather widely distributed throughout the cortex. At a more mechanistic level, Hebb (1949) and Gerard (1949) further enforced the idea that persistent reverberation of activity within neuronal networks activated at the time of encoding was likely to be crucial for converting short-term memory into long-term memory.

One of the most convincing arguments in favor of a consolidation process resides in the temporal aspects of retrograde amnesia. This form of amnesia has been typically reported to be temporally graded and characterized by a loss of recent memories, but with a relative sparing of more remote ones, just as initially observed by Ribot. It was Scoville and Milner (1957) who opened a new era in the search for the anatomical correlates of temporally graded retrograde amnesia. Patient H.M. was diagnosed with severe intractable epilepsy whose focus was located within the medial temporal lobe. Scoville, a neurosurgeon, therefore decided to excise a large portion of H.M.'s medial temporal lobe bilaterally. While effective in abolishing his epileptic seizures, this surgery unexpectedly produced a severe, temporally graded retrograde amnesia that extended back approximately 11 years (Corkin 2002). However, H.M.'s remote memories acquired during his youth remained unaffected, as did his intellectual capabilities. The memory profile exhibited by H.M. and other similar case studies was interpreted as evidence for a time-limited role of the hippocampus and related structures of the medial temporal lobe in memory storage (Rempel-Clower et al. 1996). In short, once consolidated, memories must be stored elsewhere in the brain.

Numerous animal studies were later successful in reproducing the temporally graded retrograde amnesia observed in humans. These studies particularly focused on attempting to define precisely the duration of the memory consolidation phase. Historically, the most classically used protocol for this type of study consisted in evaluating the duration of sensitivity to amnesia by applying amnestic treatments at different time periods after the initial acquisition of the information and then testing for memory retrieval. The potential for enhancing memory by applying specific drug treatments during the course of the consolidation phase further stimulated interest in this type of experiment (McGaugh and Herz 1972; McGaugh 1989). Taking advantage of the development of refined animal models of human memory in primates and rodents and of the use of innovative lesion approaches, many researchers then succeeded in identifying the anatomical components of the medial temporal lobe memory system. Brain structures that constitute this system are anatomically connected and are conserved across species such as humans, non-human primates and rodents (Amaral and Witter 1989; Lavenex and Amaral 2000; Suzuki and Amaral 2003). They include the hippocampus proper (CA1, CA3, dentate gyrus) and the subicular complex, as well as adjacent entorhinal, perirhinal and parahippocampal cortices.

It rapidly became clear that the medial temporal lobe system was specifically important for declarative memory of the kind affected in patient H.M (Milner el al. 1998). Findings from all of these studies were not only reminiscent of Müller and Pilzecker's original proposal that the consolidation process represented the progressive post-acquisition stabilization of memories, but also provided a theoretical basis to differentiale at least three forms of memory that can be distinguished on the basis of their timescales: a labile and vulnerable form, named short-term memory, that processes information from seconds to minutes; a more stabilized form, known as long-term memory, where information can be stored from minutes to several days; and an enduring form, termed remote memory, that ensures storage for extended periods of time lasting from several weeks to a lifetime (Fig. 1). As detailed in the next paragraph, the different kinetics for short-term, long-term and remote memory directly reflected on the very concept of consolidation. These have led to what is currently referred to as "synaptic consolidation", a rapid process of experience-related synaptic changes, and system consolidation, which involves a much slower and elaborated neuronal reorganization at the level of entire hippocampal-cortical neuronal circuits (McGaugh 2000).

Two main phases in memory consolidation

Neurobiologists distinguish two phases of memory consolidation on the basis of their specific temporal kinetics. Synaptic consolidation occurs within minutes to hours following a learning experience and is triggered by a series of cellular and molecular cascades that include release of neurotransmitters, activation of central receptors, production of second messengers, activation of various transcription factors and associated gene signaling pathways. Ultimately, these now well-characterized metabolic events lead to the synthesis of proteins that are directly responsible for the occurrence of morphological changes within localized neuronal networks, such as changes in synaptic strength and restructuring of existing connections as well as the growth of novel ones (Squire and Kandel 2000; Kandel 2004). Consolidation can also occur at a system level. In this case, system consolidation is a much slower process that can last for weeks to months, even up to several decades depending on the species. This process is characterized by a gradual reorganization of the brain regions, namely hippocampal and cortical circuits that support memory processing (Dudai 2004; Frankland and Bontempi 2005). The concept of system consolidation was directly inspired from studies of retrograde amnesia gradients that indicated that the hippocampus cannot be the permanent repository of long-term memory (Zola Morgan and Squire 1990). This idea

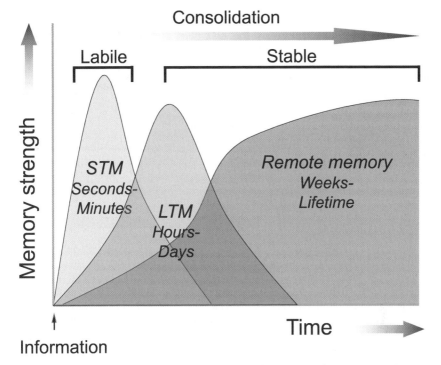

Fig. 1. Different stages in memory formation. Experimental and clinical evidence point to the existence of three main forms of independent memory that undergo parallel processing and consolidation, albeit on different temporal kinetics ranging from only seconds to several months or even a lifetime (adapted from McGaugh 2000). Consolidation is defined as the generic process enabling memory traces to gradually acquire stability over time. STM, short-term memory; LTM, long-term memory

actually forms the central tenet of most contemporary views of system consolidation: the hippocampus acts as a temporary consolidation organizing device, but storage and retrieval of enduring memories depend on broadly distributed extrahippocampal networks, presumably located in the cortex.

The standard model of memory consolidation versus the multiple trace theory: two divergent views of the same process

In the so-called standard model of memory consolidation (Fig. 2), the hippocampus is believed to rapidly integrate and bind together information transmitted from distributed cortical networks that support the various features of a whole experience in order to form a coherent memory trace. Consolidation of this new memory trace at the cortical level would then occur slowly via repeated reactivation of hippocampal-cortical networksr to progressively increase the strength and stability of cortical-cortical connections. Over time, as memories mature, the role of the hippocampus would gradually diminish, leaving extrahippocampal regions, presumably cortical areas, to become in-

Cortical modules

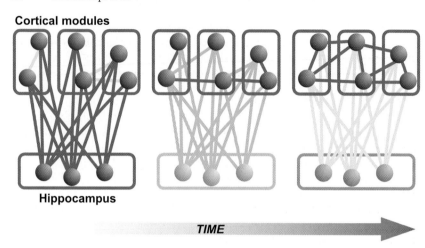

Hippocampus

TIME

Fig. 2. The standard model of memory consolidation. Perceptual, motor and cognitive information is initially processed by several specialized primary and associative cortical areas represented by cortical modules. The medial temporal lobe, including the hippocampus and related structures, integrates the various features of an experience and fuses them rapidly into a coherent memory trace. System consolidation then occurs slowly over time and involves a hippocampal-cortical dialogue to gradually strengthen cortical-cortical connections. This process involves not only the strengthening of existing cortical-cortical connections but also the creation of new connections between previously unconnected neurons via repeated activation of hippocampal-cortical networks during periods of quiet wakefulness or sleep. As cortical memories mature and acquire stability, the functional role of the hippocampus gradually diminishes, enabling structured cortical networks to ensure retrieval of remote memories independently. A key feature of this model is that changes in the strength of hippocampal-cortical connections are fast but short-lasting whereas changes within cortical-cortical connections are slow but long-lasting. (Adapted from Frankland and Bontempi 2005)

dependently capable of sustaining permanent memories and mediate their retrieval (Squire and Alvarez 1995; McClelland et al. 1995).

An alternative and challenging view is offered by the multiple trace theory, which posits that the hippocampus retains a permanent role in memory storage and retrieval as long as memories exist (Nadel and Moscovitch 1997). This view is supported by three main lines of clinical and experimental observations that cannot be accounted for by the standard model of memory consolidation. First, retrograde amnesia can, in some cases, be ungraded (i.e., 'flat'), wherein both recent and remote memories have been reported to be similarly impaired (Cipolotti et al. 2001). Second, certain retrograde amnesia gradients have been reported to last for decades or up to almost the entire human life span in some amnesic patients, thus raising the question of the ethological value of such an extended period of consolidation. Third, the observation of a retrograde amnesia gradient and its temporal extent may depend on the type of declarative memory to be consolidated (episodic, semantic or spatial; Moscovitch et al. 2006). The main features of the multiple trace theory are as follows. Each consciously experienced event would consist of a cohesive hippocampal-cortical ensemble. Each time that this particular event is recalled, it would be recreated and recoded in the form of multiple and stronger related memory traces dispersed over larger areas of hippocampal-cortical networks. Therefore, the relative sparing of remote memories in amnesic patients would be a function of the extent of hippocampal damage, with limited damage producing temporally graded retrograde amnesia and extensive lesions, resulting in a flat gradient for retrograde amnesia. Nadel and Moscovitch (1997, but for review, see also Moscovitch et al. 2006) further posit that the recurrent creation of multiple hippocampal-cortical traces would predominantly favor the integration of information with preexisting knowledge to form old semantic memories (memory for general knowledge of facts) whose retrieval could possibly occur without the contribution of the medial temporal lobe memory system. However, in the case of remotely acquired episodic memories, which are autobiographical and richly detailed in nature, these authors postulated that retrieval would always require the contribution of hippocampal-cortical networks.

It must be noted that the results of some case studies, for instance, patient E.P., who suffered large lesions of the medial temporal lobe following an episode of viral encephalitis, do not support the multiple trace theory (Teng and Squire 1999). Despite extensive hippocampal damage, this patient had excellent autobiographical memories from his youth and could accurately recall the spatial layout of the area where he grew up more than 50 years earlier. Even more astonishing was the ability of E.P. to mentally navigate and construct novel spatial routes to access specific locations of his past neighborhood, thus pointing to the existence of extrahippocampal spatial maps acquired long ago. In accordance with this case study, temporal gradients have been reported in animals even after complete hippocampal lesions. These discrepancies, especially the issue as to whether recalled memories in hippocampal patients such as E.P. or K.C. (Rosenbaum et al. 2000) are as vivid and richly detailed as in healthy patients, continue to be hotly debated (Bayley et al. 2003; Squire et al. 2004; Moscovitch et al. 2006).

Time-dependent hippocampal-cortical interactions during system consolidation as revealed by brain imaging approaches

In humans

As outlined in the brief history above, temporally graded retrograde amnesia has been reported in humans and these case studies are well documented (Squire et al. 2004; Moscovitch et al. 2006). One striking difference with animal studies concerns the length of the gradient. Most often, amnesia extends back across a period of several years and, in some cases, even several decades. Both remote semantic memory for facts and remote episodic memory for autobiographical events appear to be spared, although some patients with hippocampal damage have been diagnosed as having difficulties in recollecting detailed episodes from their early life (Rosenbaum et al. 2000), possibly because of extensive damage outside the medial temporal lobe (Squire et al. 2004). Results from positron emission tomography (PET) and functional brain imaging (fMRI) studies in humans have proven to be controversial, in part because of technical difficulties in recording activity across subjects accurately and in designing appropriate comparative baseline control tasks aimed at isolating brain activity associated with non-mnemonic aspects of the testing procedure. However, Haist and colleagues (2001) succeeded showing in the existence of a time-limited role of the medial temporal lobe in remote memory in healthy adults. Subjects were asked to identify faces of people that were famous during a particular decade from the 1940s to the 1990s. Robust hippocampal activation, as revealed by fMRI, was observed only during identification of faces that achieved fame recently, during the 1990s, whereas the entorhinal cortex, which connects the hippocampus to other widespread cortical regions, appeared to have a more prolonged role in supporting remote memory over decades (Eichenbaum 2001). A recent fMRI study by Takashima and colleagues appears to be even more consistent with the imaging data obtained in animals and the observed time course of the retrograde amnesia gradient (Takashima et al. 2006). In contrast to the study by Haist, the authors were successful in studying remote memory in a prospective manner, as in the experimental designs classically used in animals. Healthy adult subjects were asked to encode and memorize a series of photographs depicting various natural landscapes. Recognition of these photographs was then probed during a testing session that occurred the same day, and 1 day, 1 month or 3 months later. Results show a striking correspondence with the imaging results obtained in animals. Not only was confident recall of remotely acquired pictures independent of the hippocampus and dependent on the medial prefrontal cortex, but this dramatic time-dependent reorganization also occurred at a rapid rate, of a magnitude comparable to that observed in animal studies of retrograde amnesia (Frankland and Bontempi 2006).

In animals

When it comes to studying temporally graded retrograde amnesia, animal models offer a certain number of advantages over human case studies (Squire 1992). First, the content and age of memories can be rigorously assessed across time points in a prospective manner, two parameters that are difficult to control for in humans, where studies rely almost exclusively on retrospective questionnaires. Second, the location and extent of lesions are strictly circumscribed to the brain area of interest. As reviewed by Frankland and Bontempi (2005), more than 30 studies have investigated the impact of disrupting hippocampal function on the retrieval of recent and remote memory. Close analysis reveals that the length of the retrograde amnesia gradient seems to be a function of several factors, including the species, the type and complexity of the memory paradigm, the amount of training and the type, extent and location of the brain lesion. The typical pattern of memory impairment induced by hippocampal lesions as a function of the age of the memories that emerges from all these studies is summarized in Fig. 3.

Although temporally graded retrograde amnesia has been shown consistently across a variety of species, both in invertebrates (e.g., bees or flies) and vertebrates (e.g., nonhuman primates, rabbits, rodents, or pigeons), this phenomenon has been demonstrated almost exclusively by performing various lesions of the medial temporal lobe and related structures. Unfortunately, these lesion approaches do not provide any insights into the status of memory storage in the hippocampus at the time of memory retrieval. In other words, one fundamental question remains: what would have been the contribution of the hippocampus to remote memory retrieval had it been functionally intact at the time of retention testing? One powerful way to address this question is by using functional brain imaging. Over the past two decades considerable progress has

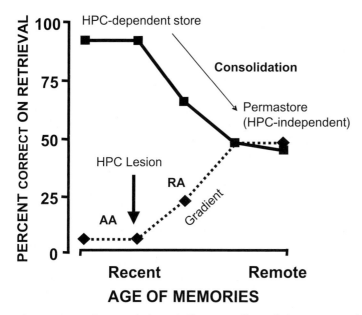

Fig. 3. Retrograde amnesia is typically temporally graded. Damage to the medial temporal lobe, including the hippocampus (HPC), impairs recent memory while sparing remote memory, pointing to a time-limited role of the hippocampus in memory storage. The period of time covered by the retrograde amnesia (RA) gradient corresponds to the duration of system consolidation that renders memory traces hippocampus-independent. Hippocampal lesions also induce incapacity to form any new enduring memories. AA, anterograde amnesia

been made and techniques such as PET, and more recently fMRI, have made it possible to visualize the human brain at work. Similar techniques are also available in small animals. One such technique is the (^{14}C)2-deoxyglucose autoradiographic method, which can be readily used in freely moving animals confronting various cognitive challenges (Sokoloff et al. 1977). Given that glucose is the sole energy source for neurons in normal physiological conditions, the measure of its rate of consumption in a given brain region enables to evaluate the level of functional implication of that region in a particular experimental situation. The use of the radioactive analogue 2-deoxyglucose, which is captured by active neurons and trapped inside them, allows to obtain quantitative assessments of regional levels of cerebral glucose utilization, thereby providing instant snapshots of changes in neuronal activity throughout the brain. We adapted this technique in mice submitted spatial discrimination testing in the eight-arm radial maze, wherein animals had to locate a set of three constantly baited arms using spatial cues scattered throughout the room containing the maze. Our main goal was to visualize memory reorganization during system consolidation in an intact (non-lesioned) brain and to determine whether retrieval of remotely acquired spatial information would be associated with a decreased functional involvement of the hippocampal region as compared to the situation where subjects were recalling recent information (Bontempi et al. 1999). Following acquisition of the task, animals were submitted to retention testing either five days (recent memory) or 25 days (remote memory) later. The choice of these retention delays for testing recent and remote memory was made on the basis of previous observations concerning the length of the retrograde amnesia gradient observed in mice following restricted lesions to the entorhinal cortex (Cho et al. 1993). Intravenous injections of 2-deoxyglucose were carried out 1 min prior to each retention session. Brains were collected immediately after the 35-min retention session and prepared for autoradiography. As shown in Fig. 4, region-specific changes in metabolic activity occur as memories matured over time. In accordance with the predictions of the standard model of memory consolidation, recall of remote as compared to recent memories was associated with a significant reduction of metabolic activity within the different regions of the hippocampal formation, including the hippocampus proper (CA1, CA3 and dentate gyrus) as well as the entorhinal cortex. This functional disengagement of the hippocampal formation was associated with increased metabolic activity in frontal, anterior cingulate and temporal cortices, indicating a greater level of recruitment of these regions in remote memory retrieval (Fig. 4). This particular profile of neuronal reorganization was specific to memories acquired 25 days earlier, since introducing a change in reward contingencies (exposure to a different set of baited arms using the same maze as initial acquisition or placing the maze in a new room with a different spatial layout) resulted in hippocampal reactivation, indicating that this region is required to encode new information and to maintain access to recently acquired information until further corticalization (Bontempi et al. 1999).

More recently, we conducted complementary imaging experiments at a more cellular level by mapping the expression of activity-dependent genes such as *c-fos* and *zif268* in mice submitted to a spatial discrimination task in a five-arm maze designed in our laboratory (Durkin et al. 2000). *c-fos* and *zif268* are two inducible immediate early genes that are required for synaptic plasticity and memory formation and whose expression is routinely used as an index of neuronal activation (Jones et al. 2001; Hall et al. 2001; Fleischmann et al. 2003). Groups of mice were trained to locate the invariant spatial position of one baited target arm in the five-arm maze (criterion of 80% correct response) and subsequently tested 1 day (recent memory) or 30 days after initial acquisition (Fig. 5A). To isolate gene expression associated with the nonmnemonic aspects of the spatial testing procedure (arousal, locomotor activity, etc.), we generated paired control mice that were exposed to the maze but were not confronted to any arm choice during initial training or retention testing. For this group, only one target arm at a time was opened and this arm was either reinforced or not as a function of the response (correct or incorrect) of the paired experimental subjects. Brains were dissected 90 min after the end of a single trial of retention testing to prevent any reacquisition processes and were prepared for immunocytochemistry (Maviel et al. 2004). Again, a converging and consistent pattern of neuronal reorganization emerged: retrieval of recent memory engaged the hippocampus whereas retrieval of remote memories was associated with increased neuronal activity in widespread cortical regions such as the prefrontal, anterior cingulate and retrosplenial cortices (Fig. 5B; Fig. 6A). Reorganization of neuronal activation also occurred at a subregional level across cortical layers. In the parietal cortex, neuronal activation actually shifted from the deep cortical layers V-VI to more superficial layers II-III and IV (not shown, but see Maviel et al. 2004). Such laminar reorganization was consistent with the formation of cortical-cortical assemblies possibly constituting major sites of storage of encoded information, including the establishment

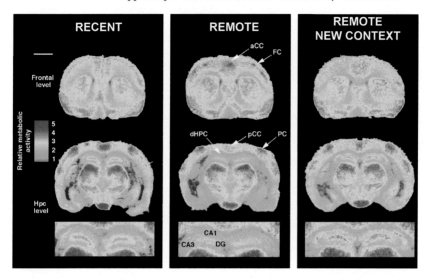

Fig. 4. System consolidation as revealed by functional brain imaging using the $(^{14}C)2$-deoxyglucose autoradiographic technique in mice. Color-coded autoradiographs of coronal sections of mice tested for either recent memory (5 days, *left*), remote memory (25 days, *center*) or remote memory in the new context (30 days, *right*) following initial training in the eight-arm radial maze.The lower section of each panel shows a magnified view of the dorsal hippocampus (dHPC) that includes the CA1, CA3 and dentate gyrus (DG) regions. Lengthening the retention interval from 5 to 25 days resulted in a significant decrease in metabolic activity in the hippocampus and a concomitant increase in activity in several cortical regions, such as the frontal (FC) and anterior cingulate (aCC) cortices. These data point to a time-limited role of the hippocampus in remote memory storage. Over time, as memories mature, distributed cortical areas become capable of mediating retrieval of remote memories independently. However, when mice are confronted with a different context at a remote time point, the hippocampus is re-engaged, indicating that this brain region is required to encode novel information. PC, parietalcortex; pCC, posterior cingulate cortex. Scale bar, 2 mm. (Adapted from Bontempi et al. 1999)

of extrahippocampal maps (Save and Poucet 2000). The parietal cortex may also constitute a storage site for very familiar environments, as recently suggested by a study using rats reared in a complex spatial environment. Despite subsequent extensive hippocampal damage, these animals were able to remember spatial representations of their environment (Winocur et al. 2005).

Overall, the recruitment of these different cortical regions may reflect their increasing roles in memory storage, effortful recall, performance monitoring and use of consolidated remote memories (Buckner et al. 1999; Markowitsch 1995; Piefke et al. 2003; Ridderinkhof et al. 2004). However, since imaging data are only correlative, we next adopted an invasive approach that consisted of transiently silencing neuronal activity in the discrete brain regions identified on the basis of our previous imaging data. This approach was chosen over classical irreversible lesions to minimize compensatory phenomena within memory systems (Bures and Buresova 1990). Lidocaine, an anesthetic that blocks sodium channels and abolishes action potentials, was infused immediately prior to each retention session in specific brain regions such as the

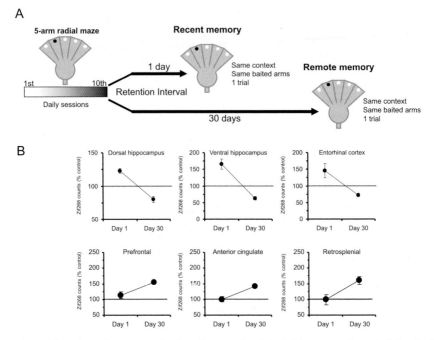

Fig. 5. Time-dependent reorganization of neuronal activity in hippocampal-cortical circuits during system consolidation. (**A**) Testing procedure in the five-arm maze. Mice were first submitted to spatial discrimination testing in the five-arm radial maze, wherein they were trained to locate and remember the spatial position of one target, constantly baited arm. Following 10 days of training to acquisition criterion (80% correct responses), memory performance was then assessed either 1 day (recent test) or 30 days (remote test) later using single retention trial to prevent reacquisition phenomena. Animals underwent intracardiac perfusions 90 min after retention testing. Their brains were collected and prepared for visible immunocytochemistry. (**B**) Cellular expression of activity-dependent genes. Zif628 counts relative to paired controls (100% baseline) are shown in the different regions of the hippocampal formation (dorsal and ventral hippocampus and interconnected entorhinal cortex) and in cortical regions (prefrontal, anterior cingulate and retrosplenial cortices) following recent (1 day) and remote (30 days) memory testing

hippocampus or one of the different cortical areas. Results corroborated the imaging data by showing that recall of recently acquired information is initially dependent on hippocampal-cortical networks (hippocampus and posterior cingulate cortex) and then becomes progressively more dependent on distributed cortical-cortical networks (prefrontal and anterior cingulate cortices; Maviel et al. 2004). A similar pattern of time-dependent reorganization among hippocampal and cortical regions was observed in animals processing recent and remote contextual fear memories (Frankland et al. 2004).

What are the neuronal mechanisms involved in remote memory storage? While multidisciplinary approaches in animals have contributed to the elucidation of many of the cellular and molecular mechanisms underlying memory processing in hippocampal networks during synaptic consolidation, much less is known about how the dynamics of hippocampal-cortical interactions allows these memories to be transformed into

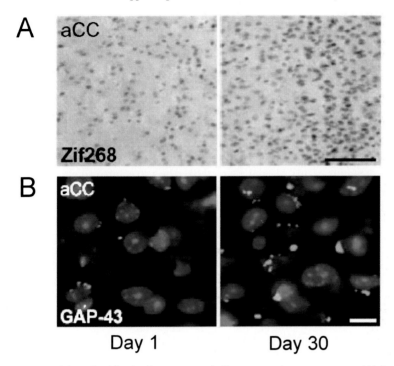

Fig. 6. Wiring plasticity in the cortex underlies remote memory storage. (A) Representative photomicrographs showing increased Zif268 staining in the anterior cingulate cortex (aCC) on day 30 as compared to day 1. (B) The expression of growth-associated protein 43 (GAP-43), a presynaptic protein that controls axon growth and sprouting and is considered to be a marker of newly formed synapses, was examined in the aCC of animals tested either 1 day (recent memory) or 30 days (remote memory) after spatial discrimination in a five-arm maze. Cortical consolidation of the spatial position of the baited arm of the maze was accompanied by an increase in GAP-43 staining on day 30 as compared with day 1. Nuclei appear in blue (DAPI staining), fluorescent GAP-43 staining is green. This finding supports the concept that cortical networks undergo structural changes as memories mature over time. The addition of novel synapses may have contributed, at least in part, to increasing the extent and complexity of neuronal networks, as shown by the increased number of Zif268-positive nuclei in the aCC. *Scale bars*: (A) 100 μm, (B) 10 μm. (Adapted from Maviel et al. 2004)

remote memories in cortical networks during the course of system consolidation. Memory reactivation remains the core mechanism in most models of system consolidation. Repeated activation of hippocampal networks is thought to reinstate, in a coordinated and synchronized manner, activity in different cortical networks. Reactivation can be induced either during "online" states of conscious recollection or quiet wakefulness or during awake periods immediately after spatial experience (Foster and Wilson 2006). Replay of neuronal patterns of activity associated with a learning experience also happens during "offline" states, such as sleep. Marr (1971) was among the first to suggest that offline coordinated replay of previous waking patterns of neural activity during sleep may drive consolidation and ensure the gradual stabilization of cortical memory traces. A role for sleep in memory consolidation is now well documented. Although exceptions exist, a gain in retention performance after brief naps or overnight sleep has been reported (Hoffman and McNaugton 2002; Mednick et al. 2002; Stickgold et al. 2000; Wagner et al. 2004; Walker et al. 2002). Conversely, sleep deprivation can interfere and even block memory consolidation. Particularly interesting is the observation that beneficial effects of sleep appear to be dependent on the form of memory. While consolidation enhancement of procedural and motor learning have been consistently reported, the data are less clear with regard to declarative memory, raising the possibility that memory consolidation may involve several steps, each being driven by specific phases of sleep such as slow-wave sleep (SWS) or rapid eye movement sleep (REM), during which dreaming occurs. Sleep is characterized by a structured combination of neuronal oscillation that is region-specific. In the hippocampus, SWS exhibits high-frequency network oscillations known as "ripples," whereas neocortical SWS activity is organized into low-frequency oscillations referred to as spindles. Importantly, hippocampal ripple activity has been shown to occur in temporal correlation with cortical spindle activity. Such coordinated replay of experience-dependent activity could thus represent one important mechanism that drives not only stabilization and refinement of memory traces in the cortex but also the updating of pre-existing cortical memories (Siapas and Wilson 1998; Ribeiro and Nicolelis 2004). Temporally structured replay of waking patterns of hippocampal ensemble activity associated with earlier learning is also subsequently observed during REM sleep, supporting the possibility that this particular phase of sleep contributes to memory trace reactivation (Louie and Wilson 2001; Peigneux et al. 2004).

At the molecular level, sleep has been associated with the up-regulation of around 100 known genes (and about 400 as yet unidentified ones), independently of the phases of the circadian rhythm (Cirelli 2004). It is reasonable to assume that at least some of these genes, either directly or indirectly via triggering of specific signaling pathways, mediate the changes in synaptic plasticity within distributed cortical networks. Two main forms of plasticity are thought to coexist within hippocampal-cortical networks (Chklovskii et al. 2004). In the hippocampus, rapid encoding predominantly involves "weight plasticity," that is changes in the strength between existing connections via long-term potentiation (LTP) and long-term depression (LTD) mechanisms that modulate synaptic efficacy (Bliss and Collingridge 1993; Frey and Morris 1998; Laroche et al. 2000). In the cortex, however, where network reorganization occurs much more slowly, an additional form of plasticity may involve the creation of novel synapses between previously unconnected neurons, a process termed synaptogenesis (Benowitz and Routtenberg 1997). It is this form of "wiring plasticity" that we observed in our brain imaging studies in mice (Fig. 6B). Long-term storage of spatial and contextual information was accompanied by the synthesis of new synapses in specific areas of the prefrontal and anterior cingulate cortices, thus raising the possibility that wiring plasticity contributed significantly to changing the organization and architecture of distributed cortical networks supporting remote memory storage (Frankland et al. 2004; Maviel et al. 2004). Such morphological changes appear to be crucial, as α-calcium/calmodulin-dependent protein kinase II mutant mice exhibiting normal hippocampal plasticity, but lacking cortical wiring plasticity, are unable to form enduring memories (Frankland et al. 2001, 2004). Since plasticity phenomena occur at different rates in hippocampal and cortical networks, the hippocampus has been conceptualized as a fast learning area whereas the cortex would be a slow learner (McClelland et al. 1995). Although the mechanisms supporting this apparent division of labor are still unclear, it could help prevent catastrophic interference as emphasized in current connectionist models, protect existing cortical memories from being erased and ensure constant updating of these cortical memories.

A privileged role for the prefrontal cortex in the processing of remote memory

Our imaging studies, coupled to region-specific inactivations, identified the prefrontal cortex as playing a privileged role in processing remote memory. Classically, the prefrontal lobe has been implicated in mediating temporal sequencing, effortful monitoring and response selection during memory retrieval (Ungerleider 1995; Ridderinkhof et al. 2004). Forgetting, which may reflect a partially degraded memory trace, is inevitably associated with remote memory storage; in our imaging studies, it is conceivable that mice had to deploy more effort to access and retrieve remotely acquired as compared to recently acquired information. Our findings point to additional roles played by the prefrontal cortex. Anatomically, the prefrontal cortex consists of several highly interconnected regions, including the prelimbic, infralimbic and anterior cingulate areas. These regions are reciprocally connected to sensory, motor and limbic cortices (Uylings et al. 2003) and are therefore ideally placed to integrate complex information from various sensory sources. This potential for integration has led us to suggest that the ability of the prefrontal cortex to process remote memories might mirror that of the hippocampus in the processing of recent information. At the time of encoding, the hippocampus is thought to integrate information from various independent cortical modules that represent the various features of an experience and then to rapidly fuse these features into a coherent memory trace (Eichenbaum 2004). As memories become gradually consolidated in the cortex through strengthening of cortical-cortical connections, this integrative function might be transferred to the prefrontal cortex, which would then become capable of integrating information embedded within other distributed cortical networks (Fig. 7). The prefrontal cortex may play another important executive role, that is, the modulation of hippocampal activation during remote memory recall. Closer examination of our imaging data revealed that neuronal activity in the hippocampus at the time of remote memory recall was actually significantly lower than that of control subjects exposed to the maze environment (Fig. 5B; Frankland et al. 2004; Maviel et al. 2004). This observation has two implications. First, it indicates that the observed reduction in hippocampal activation over time is not solely the consequence of a reduction in size of the neuronal population representing the familiar (consolidated) memory or of a more widely distributed hippocampal networks supporting memory storage. Second, it raises the possibility that inhibitory influences arising from other brain regions, presumably cortical, that are actively engaged in memory storage and retrieval may progressively control the level of hippocampal participation during the course of memory consolidation. The prefrontal lobe has been shown to play a critical role in executive functions via a top-down control over posterior association cortical regions involved in sensory processing and voluntary recall (Tomita et al. 1999). We sought to examine whether similar executive control could be exerted over the hippocampus at the time of remote memory retrieval. In preliminary experiments involving the five-arm maze, inactivation of the prefrontal cortex at the 30-day retention interval resulted in a marked reactivation of the hippocampus. This effect was specific to the prefrontal cortex as it did not occur following inactivation of the other targeted cortical areas such as the anterior cingulate cortex. Functionally, the prefrontal cortex might control hippocampal activity via direct anatomical connections with the hippocampus or indirect modulation of posterior cortical areas from which originate entorhinal-hippocampal projections. A major role for this top-down inhibitory control would be to prevent encoding of existing (already corticalized), and therefore redundant, information (Fig. 7). In this scenario, whether or not inhibition occurs would be a function of the status of pre-existing knowledge already stored in the cortex. If retrieval is successful, hippocampal activation will rapidly be inhibited. However, in the case of a mismatch situation, for example, if the actual situation requires encoding of novel information or if the information that is to be retrieved has been partially or totally forgotten, this inhibition will not be exerted and the hippocampus will be reengaged to enable updating and reconsolidation (Dudai 2004, 2006). This executive, top-down modulation of hippocampal activity is compatible with the existence of comparator cells in the prefrontal cortex whose function is to evaluate the occurrence of match versus mismatch situations.

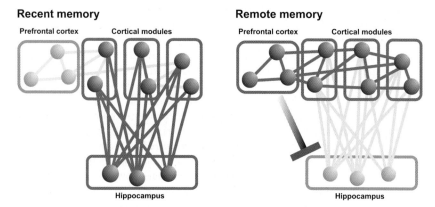

Fig. 7. A privileged role for the prefrontal cortex (PFC) in system consolidation. At the time of encoding, the hippocampus is crucial in integrating information from distributed cortical modules, each representing individual components of a memory. As memories mature over time, cortical-cortical connections are strengthened, allowing cortical regions, and particularly the PFC, to ensure retrieval of remote memories independently and to assume the integrative role initially exerted by the hippocampus. Consistent with this model, inactivation of the PFC blocks expression of remote memory but spares recent memory in rodent tests of contextual and spatial memory. The PFC might also exert executive control upon hippocampal activity during remote memory recall. The hippocampus is normally active when processing the outside environment. However, when the incoming information matches previously stored cortical information, the PFC will inhibit hippocampal activity via direct or indirect efferent pathways to prevent encoding of redundant information. In the presence of a mismatch situation, no such inhibition is exerted and the hippocampus will be engaged as usual. (Adapted from Frankland and Bontempi 2005)

Conclusion

The definition of the term "consolidation" that is found in any dictionary is simple: to strengthen or to make firm. Achieving consensus about its neurophysiological correlates, however, has proven to be much more challenging, and despite its initial conceptualization more than a century ago, the concept and the neuronal mechanisms underlying the consolidation process continue to be hotly debated in the field of cognitive neuroscience.

One must acknowledge that, over the years, considerable progress has been made in unraveling the neurobiological bases of memory consolidation. Functional brain imaging techniques in particular, along with genetic approaches in mice and traditional pharmacological and lesion approaches, have provided new breakthroughs in the identification of the different brain regions and neuronal networks supporting recent and remote memory storage. The consolidation process is no longer seen as a passive phenomenon but as a dynamic process of neuronal reorganization that affects extensive hippocampal and cortical networks over time scales that are much longer than those initially envisioned. While in accordance with the standard model of memory consolidation, our own findings highlight the fact that the cortex does not simply serve as a passive storage site but is also capable of actively integrating new memories, depending on the organization and status of pre-existing knowledge. The prefrontal cortex in particular appears to play a crucial role in integrating information from distributed cortical networks and modulating hippocampal activity during memory recall. It is our view that the maturation and consolidation of cortical memory traces should not be considered as a one time process but rather seen as an "open-ended" process involving continuous indexation and reorganization of pre-existing cortically stored information.

However, despite considerable advances in the field of functional genomics and proteomics, we are still far from understanding all the precise cellular and molecular mechanisms underlying memory reorganization within cortical networks during the course of system consolidation. Numerous questions remain to be addressed. For instance, given the ongoing debate in the human memory field (Moscovitch et al. 2006), one might wonder whether the same pattern of neuronal reorganization over time applies equally to semantic and episodic declarative memories and with similar rates in each case. Does the eventual corticalization of memory traces entail a change in the initial format so that there would be an inevitable loss of details and richness? Experiments in intact animals using brain imaging approaches should contribute to resolving these debates but will require adequate tasks that effectively dissociate semantic from episodic components. Another question concerns the different kinetics of memory processing in the hippocampus compared to the cortex. Storage in the cortex happens slowly, presumably to prevent catastrophic interference and erasure of existing memories. In contrast, plasticity in the hippocampus can be modified at a fast pace. It is conceivable that different plasticity mechanisms coexist within these brain regions, but their cellular and molecular correlates remain to be elucidated. Interestingly, upon reactivation, cortical memories have been shown to become labile again and to require another round of protein synthesis to regain stability. While this process, termed reconsolidation, highlights the possibility for updating existing stored cortical memories, the similarity and the potential differences in the mechanisms governing consolidation and reconsolidation are yet to be identified (Dudai 2006).

Answers to these questions will require integrative approaches that involve cross-species analyses and combine several levels of analyses ranging from innovative cognitive paradigms to cellular, molecular, imaging and electrophysiological techniques, enabling us to probe not only the functional involvement of single neurons but also the functioning of complex neuronal ensembles.

Acknowledgements. Support was provided by grants from the CNRS (CNRS UMR 5106), the Action Concertée Incitative "Temps et Cerveau" and the Fédération pour la Recherche sur le Cerveau (FRC). We would like to thank our collaborators who contributed significantly to the functional brain imaging experiments in mice described here, in particular Catherine Laurent-Demir, Claude Destrade, Robert Jaffard, Thibault Maviel and Frédérique Menzaghi.

References

Amaral DG, Witter MP (1989) The three-dimensional organization of the hippocampal formation: a review of anatomical data. Neuroscience 31:571–591

Bayley PJ, Hopkins RO, Squire LR (2003) Successful recollection of remote autobiographical memories by amnesic patients with medial temporal lobe lesions. Neuron 38:135–144

Benowitz LI, Routtenberg A (1997) GAP-43: an intrinsic determinant of neuronal development and plasticity. Trends Neurosci 20:84–91

Bliss TVP, Collingridge GL (1993) A synaptic model of memory: long-term potentiation in the hippocampus. Nature 361:31–39

Bontempi B, Laurent-Demir C, Destrade C, Jaffard R (1999) Time-dependent reorganization of brain circuitry underlying long-term memory storage. Nature 400:671–675

Buckner RL, Kelley WM, Petersen SE (1999) Frontal cortex contributes to human memory formation. Nature Neurosci 2:311–314

Bures J, Buresova O (1990) Reversible lesions allow reinterpretation of system level studies of brain mechanisms of behavior. Concepts Neurosci 1:69–89

Burnham WH (1903) Retroactive amnesia: illustrative cases and a tentative explanation. Am J Psychol 14:382–396

Chklovskii DB, Mel BW, Svoboda K (2004) Cortical rewiring and information storage. Nature 431:782–788

Cho YH, Beracochea D, Jaffard R (1993) Extended temporal gradient for the retrograde and anterograde amnesia produced by ibotenate entorhinal cortex lesions in mice. J Neurosci 13:1759–1766

Cipolotti L, Shallice T, Chan D, Fox N, Scahill R, Harrison G, Stevens J, Rudge P (2001) Long-term retrograde amnesia...the crucial role of the hippocampus. Neuropsychologia 39:151–172

Cirelli C, Gutierrez CM, Tononi G (2004) Extensive and divergent effects of sleep and wakefulness on brain gene expression. Neuron 41:35–43

Corkin S (2002) What's new with the amnesic patient H.M.? Nature Rev Neurosci 3:153–160

Dudai Y (2004) The neurobiology of consolidations, or, how stable is the engram? Annu Rev Psychol 55:51–86

Dudai Y (2006) Reconsolidation: the advantage of being refocused. Curr Opin Neurobiol 16:174–178

Durkin TP, Beaufort C, Leblond L, Maviel T (2000) A 5-arm maze enables parallel measures of sustained visuo-spatial attention and spatial working memory in mice. Behav Brain Res 116:39–53

Eichenbaum H (2000) A cortical-hippocampal system for declarative memory. Nature Rev Neurosci 1:41–50

Eichenbaum H (2001) The long and winding road to memory consolidation. Nature Neurosci 4:1057–1058

Eichenbaum H (2004) Hippocampus: cognitive processes and neural representations that underlie declarative memory. Neuron 44:109–120

Fleischmann A, Hvalby O, Jensen V, Strekalova T, Zacher C, Layer LE, Kvello A, Reschke M, Spanagel R, Sprengel R, Wagner EF, Gass P (2003) Impaired long-term memory and NR2A-type NMDA receptor-dependent synaptic plasticity in mice lacking c-Fos in the CNS. J Neurosci 23:9116–9122

Foster DJ, Wilson MA (2006) Reverse replay of behavioural sequences in hippocampal place cells during the awake state. Nature 440:680–683

Frankland PW, Bontempi B (2005) The organization of recent and remote memories. Nature Rev Neurosci 6:119–130

Frankland PW, Bontempi B (2006) Fast track to the medial prefrontal cortex. Proc Natl Acad Sci USA 103:509–510

Frankland PW, O'Brien C, Ohno M, Kirkwood A, Silva AJ (2001) Alpha-CaMKII-dependent plasticity in the cortex is required for permanent memory. Nature 411:309–313

Frankland PW, Bontempi B, Talton LE, Kaczmarek L, Silva AJ (2004) The involvement of the anterior cingulate cortex in remote contextual fear memory. Science 304:881–883

Frey U, Morris RG (1998). Synaptic tagging: implications for late maintenance of hippocampal long-term potentiation. Trends Neurosci. 21, 181–188

Gerard RW (1949) Physiology and psychiatry. Am J Psychiat 106:161–173

Haist F, Bowden Gore J, Mao H (2001) Consolidation of human memory over decades revealed by functional magnetic resonance imaging. Nature Neurosci 4:1139–1145

Hall J, Thomas KL, Everitt BJ (2001) Cellular imaging of zif268 expression in the hippocampus and amygdala during contextual and cued fear memory retrieval: selective activation of hippocampal CA1 neurons during the recall of contextual memories. J Neurosci 21:2186–2193.

Hebb DO (1949) The organization of behavior: a neuropsychological theory. Wiley, New York.

Hoffman KL, McNaughton BL (2002) Coordinated reactivation of distributed memory traces in primate neocortex. Science 297:2070–2073

Jones MW, Errington ML, French PJ, Fine A, Bliss TV, Garel S, Charnay P, Bozon B, Laroche S, Davis S (2001) A requirement for the immediate early gene Zif268 in the expression of late LTP and long-term memories. Nature Neurosci. 4:289–296

Kandel ER (2004) The molecular biology of memory storage: a dialog between genes and synapses. Biosci Rep 24:475–522

Laroche S, Davis S, Jay TM (2000) Plasticity at hippocampal to prefrontal cortex synapses: dual roles in working memory and consolidation. Hippocampus 10:438–446

Lashley KS (1950) In search of the engram. Symp Soc Exp Biol 4:454–482

Lavenex P, Amaral DG (2000) Hippocampal-neocortical interaction: a hierarchy of associativity. Hippocampus 10:420–430

Lechner HA, Squire LR, Byrne JH (1999) 100 years of consolidation–remembering Muller and Pilzecker. Learn Mem 6:77–687

Louie K, Wilson MA (2001) Temporally structured replay of awake hippocampal ensemble activity during rapid eye movement sleep. Neuron 29:145–156

Markowitsch HJ (1995) Which brain regions are critically involved in the retrieval of old episodic memory? Brain Res Rev 21:117–127

Marr D (1971) Simple memory: a theory for archicortex. Philos Trans R Soc London B Biol Sci 262:23–81

Maviel T, Durkin TP, Menzaghi F, Bontempi B (2004) Sites of neocortical reorganization critical for remote spatial memory. Science 305:96–99

McClelland JL, McNaughton BL, O'Reilly RC (1995) Why there are complementary learning systems in the hippocampus and neocortex: Insights from the successes and failures of connectionist models of learning and memory. Psychol Rev 102:419–457

McGaugh JL (1989) Dissociating learning and performance: drug and hormone enhancement of memory storage. Brain Res Bull 23:339–345

McGaugh JL (2000) Memory–a century of consolidation. Science 287:248–251

McGaugh JL, Herz MJ (1972). Memory consolidation. Albion, San Francisco

Mednick SC, Nakayama K, Cantero JL, Atienza M, Levin AA, Pathak N, Stickgold R (2002) The restorative effect of naps on perceptual deterioration. Nature Neurosci 5:677–681

Milner B, Squire LR, Kandel ER (1998) Cognitive neuroscience and the study of memory. Neuron 20:445–468

Moscovitch M, Rosenbaum RS, Gilboa A, Addis DR, Westmacott R, Grady C, McAndrews MP, Levine B, Black S, Winocur G, Nadel L. (2006) Functional neuroanatomy of remote episodic, semantic and spatial memory: a unified account based on multiple trace theory. J Anat 207:35–66

Müller GE, Pilzecker A (1900) Experimentelle beiträge zur lehre von gedächtnis. Zeitschrift für Psychologie, Ergänzungsband 1, 1–300. Experimental contributions to the theory of memory. Summarized by W. McDougall (1901) in Mind 10:388–394

Nadel L, Moscovitch M (1997) Memory consolidation, retrograde amnesia and the hippocampal complex. Curr Opin Neurobiol 7:217–227

Peigneux P, Laureys S, Fuchs S, Collette F, Perrin F, Reggers J, Phillips C, Degueldre C, Del Fiore G, Aerts J, Luxen A, Maquet P (2004) Are spatial memories strengthened in the human hippocampus during slow wave sleep? Neuron 44:535–545

Piefke M, Weiss PH, Zilles K, Markowitsch HJ, Fink GR (2003) Differential remoteness and emotional tone modulate the neural correlates of autobiographical memory. Brain 126:650–668

Rempel-Clower NL, Zola SM, Squire LR, Amaral DG (1996) Three cases of enduring memory impairment after bilateral damage limited to the hippocampal formation. J Neurosci 16:5233–5255

Ribeiro S, Nicolelis MAL (2004) Reverberation, storage, and postsynaptic propagation of memories during sleep. Learn Mem 11:686–696

Ribot T (1881) Les maladies de la mémoire. Germer Baillère, Paris

Ridderinkhof KR, Ullsperger M, Crone EA, Nieuwenhuis S (2004) The role of the medial frontal cortex in cognitive control. Science 306:443–447

Rosenbaum RS, Priselac S, Kohler S, Black SE, Gao F, Nadel L, Moscovitch M (2000) Remote spatial memory in an amnesic person with extensive bilateral hippocampal lesions. Nature Neurosci. 3:1044–1048

Save E, Poucet B (2000) Hippocampal-parietal cortical interactions in spatial cognition. Hippocampus 10:491–499

Scoville WB, Milner B (1957) Loss of recent memory after bilateral hippocampal lesions. J Neurol Neurosurg Psych 20:11–21

Siapas AG, Wilson MA (1998) Coordinated interactions between hippocampal ripples and cortical spindles during slow-wave sleep. Neuron 21:1123–1128

Sokoloff L, Reivich M, Kennedy C, Des Rosiers MH, Patlak CS, Pettigrew KD, Sakurada O, Shimohara M (1977) The ^{14}C-deoxyglucose method for measurement of local cerebral glucose utilization: theory, procedure and normal values in the conscious and anesthetized albino rat. J Neurochem 28:897–916

Squire LR (1992) Memory and the hippocampus: a synthesis from findings with rats, monkeys, and humans. Psychol Rev 99:195–231

Squire LR, Alvarez P (1995) Retrograde amnesia and memory consolidation: a neurobiological perspective. Curr Opin Neurobiol 5:169–177

Squire LR, Kandel ER (2000) Memory: from mind to molecules. W.H. Freeman & Co, New York

Squire LR, Stark CEL, Clark RE (2004) The medial temporal lobe. Annu Rev Neurosci 27:279–306

Stickgold R, James L, Hobson JA (2000) Visual discrimination learning requires sleep after training. Nature Neurosci 3:1237–1238

Suzuki WA, Amaral DG (2003) Where are the perirhinal and parahippocampal cortices? A historical overview of the nomenclature and boundaries applied to the primate medial temporal lobe. Neuroscience 120:893–906

Takashima A, Petersson KM, Rutters F, Tendolkar I, Jensen O, Zwarts MJ, McNaughton BL, Fernandez G (2006). Declarative memory consolidation in humans: a prospective functional magnetic resonance imaging study. Proc Natl Acad Sci. USA 103:756–761

Teng E, Squire LR (1999) Memory for places learned long ago is intact after hippocampal damage. Nature 400:675–677

Tomita H, Ohbayashi M, Nakahara K, Hasegawa I, Miyashita Y (1999) Top-down signal from prefrontal cortex in executive control of memory retrieval. Nature 401:699–703

Ungerleider LG (1995) Functional brain imaging studies of cortical mechanisms for memory. Science 270:769–775

Uylings HB, Groenewegen HJ, Kolb B (2003) Do rats have a prefrontal cortex? Behav Brain Res 146:3–17

Wagner U, Gais S, Haider H, Verleger R, Born J (2004) Sleep inspires insight.Nature 427:352–355

Walker MP, Brakefield T, Morgan A, Hobson JA, Stickgold R (2002) Practice with sleep makes perfect: sleep-dependent motor skill learning. Neuron 35:205–211

Winocur G, Moscovitch M, Fogel S, Rosenbaum RS, Sekeres M (2005) Preserved spatial memory after hippocampal lesions: effects of extensive experience in a complex environment. Nature Neurosci 8:273–275

Woodworth RS (1929) Psychology. Henry Holt, New York

Zola-Morgan S, Squire LR (1990) The primate hippocampal formation: evidence for a time-limited role in memory storage. Science 250:288–290

From Molecule to Memory System: Genetic Analyses in Drosophila

Guillaume Isabel[1], *Daniel Comas*[2], and *Thomas Preat*[1]

Summary

Despite the fact that the *Drosophila* brain has only 100 000 cells, it produces complex behaviors and sustains various forms of learning and memory. We show here how the power of *Drosophila* molecular genetics permits the interconnecting of the different levels of memory analyses: molecular, network and system. In particular new, tools allow a precise spatial and temporal control of network activity, as well as recording of brain activity. Recent results indicate that the *Drosophila* brain is the site of complex phenomena more frequently associated with mammalian species.

Introduction

Drosophila melanogaster is one of the most intensively studied organisms in biology, and it serves as a model system for the investigation of many cellular, developmental and behavioral processes common to other species, including humans. This usefulness holds in particular for brain studies, as the *Drosophila* central nervous system (Fig. 1) is made of neurons and glia that operate on the same fundamental principles as their mammalian counterparts. Thus, most neurotransmitters are identical in flies and humans, and despite the fact that the *Drosophila* brain has only 100 000 cells (Shimada et al. 2005), it produces complex behaviors and sustains various forms of learning and memory. Beside studies of fundamental brain properties, fly models are being developed for a variety of neurodegenerative disorders, and the field is beginning to harness the power of *Drosophila* genetics to dissect pathways of disease pathogenesis and identify potential targets for therapeutic intervention. This approach is made possible by the fact that about 50% of human genes have a *Drosophila* orthologue (Rubin et al. 2000). In addition, transgenic strategies that allow the introduction of human genes into *Drosophila* continue to expand the list of modeled diseases, which now includes Parkinson's disease, Alzheimer's disease, Huntington's disease and several spinocerebellar ataxias (for reviews, see Shulman et al. 2003; Bier 2005).

Learning and memory genes

In *Drosophila*, a single associative-learning trial (called the "short protocol"), consisting of an odor (the conditioned stimulus, CS) accompanied by electric shocks (the

---

[1] Gènes et Dynamique des Systèmes de Mémoire, CNRS UMR 7637, École Supérieure de Physique et Chimie Industrielle, 10 rue Vauquelin, 75005 Paris, France
thomas.preat@espci.fr
[2] Present address: Glato Smith Kline, 25 Avenue du Quebec, 91195 Les Ulis, France

Bontempi et al.
Memories: Molecules and Circuits
© Springer-Verlag Berlin Heidelberg 2007

unconditioned stimulus, US) induces olfactory learning and memory (Quinn et al. 1974). The discovery of most of the learning and memory mutants was performed with this protocol. Multiple, spaced repetitions of this single protocol (called the "long protocol") generate long-term memory (LTM; Tully et al. 1994). We do not intend here to review in detail all *Drosophila* learning and memory genes or the various conditioning protocols (for reviews, see Waddell and Quinn 2001; Dubnau et al. 2003a; Heisenberg 2003; Davis 2004). Rather, we would like to show how recent genetic technologies greatly facilitate the identification and study of learning and memory genes.

Chemical mutagenesis versus transposable element mutagenesis

Initially, *Drosophila* memory mutants were issued from random chemical mutagenesis using ethyl methane sulfonate (EMS), a potent mutagen. But because EMS induces mostly subtle DNA alterations, the identification of the mutated gene is often difficult. The first step consists in genetically mapping the mutation by a complementation test, using *Drosophila*'s deletion collection. However, this map only defines a broad area covering many other genes. In contrast, the use of a marked, transposable P-element allows one to easily identify the gene of interest.

In the first behavioral screen, about 4 000 fly stocks carrying EMS-induced mutations were tested for their ability to learn in an olfactory conditioning assay (Quinn et al., 1974, 1979). The first *Drosophila* learning and/or memory mutants were *dunce* (*dnc*) and *rutabaga* (*rut*). Biochemical defects were observed in *rut* (reduced cAMP level) and *dnc* (increased cAMP level; Livingstone et al. 1984). Nevertheless, the cloning of the responsible gene took years and was facilitated by the use of new P-induced alleles. For example, the description of P-element insertions within 200 nucleotides of where the *rut* transcription started identified *rut* as the structural gene for the Ca^{2+}/Calmodulin-responsive adenylate cyclase (Han et al. 1992; Levin et al. 1992). New EMS-induced *dnc* mutants came from a screen for female sterility mutants (Mohler 1977). The *dnc* gene was identified by recombinational mapping of *dnc* mutations with restriction site polymorphisms as genetic markers (Davis and Davidson 1984). The first *amnesiac* (*amn*) mutant was identified in a screen for flies with affected memory (Quinn et al. 1979). However, the *amn* gene was cloned as a second site suppressor of the *dnc* female sterility phenotype from a P-element-induced allele and has been repeatedly isolated since (Feany and Quinn 1995; Moore et al. 1998; Toba et al. 1999). The *amn* encodes a putative adenylate cyclase activating peptide.

Another learning and/or memory mutant is *radish* (*rsh*; Folkers et al. 1993). The *rsh* gene was localized within a 180-kb interval in the 11D-E region of the X-chromosome, and several candidate genes were identified (Folkers et al. 1993). Recently, Chiang and colleagues reported (2004) that the responsible gene for the *rsh* phenotype was a phospholipase A2 (Chiang et al. 2004). However, this original finding failed to be reproduced by Dubnav and colleagues, and a second team has reported that *rsh* encodes a novel protein with possible nuclear localization motifs (Folkers et al. 2006). This discrepancy illustrates the difficulty linked to working with EMS-induced behavioral mutants.

P-element-based behavioral screens for learning and memory mutants have been performed. Various mutants were issued from these mutageneses, including *nalyot* (a myb-related Adf1 transcription factor; Boynton and Tully 1992; DeZazzo et al. 2000), *leonardo* (a zeta isoform of the 14-3-3 protein; Skoulakis and Davis 1996), and *volado* (two splice variants of an α-integrin; Grotewiel et al. 1998).

Dubnau and colleagues (2003b) recently performed a behavioral screen for LTM mutants, in parallel with microarray experiments, aimed at selecting genes with altered expression after LTM training. This work led to the identification of proteins involved in mRNA processing and translation (Dubnau et al. 2003b).

We recently described *crammer* (*cer*), a gene involved specifically in the set up of LTM (Comas et al. 2004). The P[GAL4] *cer* strain has a reduced LTM but a normal short-term and middle-term memory. Interestingly, in the wild-type strain, *cer* is transiently underexpressed three hours after LTM training. As the Cer peptide is an inhibitor of cysteine proteinases, the decrease in its expression shortly after intensive training must lead to a transient activation of its cysteine proteinase(s) target(s) (Comas et al. 2004). All together, these works demonstrate the power of the forward-genetic approach: learning or memory mutants are first characterized without knowing in advance the molecular function of the gene involved.

Spatial control of gene expression

A powerful tool of spatial control of gene expression is the GAL4/UAS enhancer-trap technique (Bellen et al. 1989; Wilson et al. 1989; Brand and Perrimon 1993; Fig. 2), which enables a selective activation of any cloned gene in a wide variety of tissues and cells (Fig. 2). An enhancer-trap element is a transposon containing an exogenous gene, such as the yeast transcriptional activator GAL4. Insertions that occur in close proximity to a transcriptional enhancer cause the GAL4 gene to be expressed in a pattern reflecting the enhancer's spatio-temporal regulatory properties. Thousands of lines carrying an insertion of the P[GAL4] element are available. With a single cross, flies are created that carry a second P-element with a reporter gene inserted downstream of GAL4 binding sites: the upstream activating sequences (UAS). The reporter gene is expressed in the same cells as GAL4. This flexible system allows for, for example, the labeling of a particular group of cells if the reporter gene encodes the green fluorescent protein (GFP; Fig. 2). Yang and colleagues (1995) characterized the intrinsic cells of the *Drosophila* olfactory memory center, the mushroom bodies (MBs; Fig. 1; Yang et al. 1995). Rather than being homogeneous, the 5 000 MB neurons were compound neuropils in which parallel subcomponents exhibited discrete patterns of gene expression (Crittenden et al. 1998; Strausfeld et al. 2003). Parallel channels of information flow, perhaps with different computational properties, subserve different roles (see below).

Using the enhancer-trap system, Ito and colleagues (1998) searched for MB extrinsic neurons. They used a reporter construct encoding the presynaptic protein, neuronal synaptobrevin-green fluorescent protein, and showed that output MB neurons are scarce and that, surprisingly, few MB extrinsic neurons project to the deutocerebrum, the premotor pathway that is the immediate modifier of behavior.

Temporal and spatial control of gene expression

When a gene is mutated, the biochemical function of the protein may be affected, or the transcription level of the normal mRNA may be decreased. The resulting biochemical defect may occur during development and/or adult life. Thus, a learning and/or memory

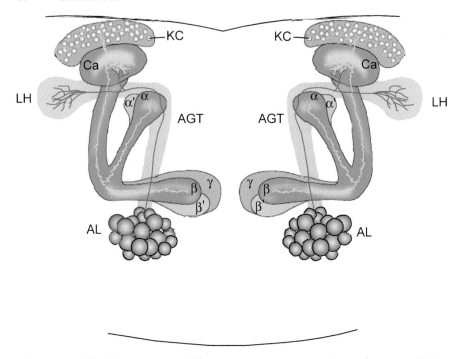

Fig. 1. *Drosophila* olfactory system. Olfactory sensory neurons project to the antennal lobes (AL). From there, projection neurons project through the antenno-glomerular tract (AGT) and connect MB dendrites localized in the calyx (Ca), as well as the lateral horn (LH). Each MB is composed of about 2,500 neurons, the Kenyon cells (KC). Three type of KC project in five lobes: α/β, α'/β' and γ

defect may be due to structural brain defects rather than to a specific physiological alteration. This issue is crucial, as even a faint developmental defect may have major consequences on brain function.

Two recent techniques, developed by the laboratory of Ronald Davis, permit the discrimination of a role in brain development from a role in memory formation *per se*. TARGET and Gene-Switch (Leung and Waddell 2004) are variants of the invaluable GAL4/UAS expression system. TARGET, which stands for Temporal and Regional Gene Expression Targeting, uses a modified temperature-sensitive yeast, GAL80 transcription factor (GAL80^{ts}) to repress GAL4 activity (Fig. 3). At low temperature (typically 18 °C), the functional GAL80^{ts} protein inhibits GAL4 transcriptional activity. At 30 °C, GAL80^{ts} becomes inactive and allows the GAL4-mediated transcription (Fig. 3). This tool can be used to rescue the defect of a memory mutant by expressing the normal protein in specific regions of the adult brain while preventing its expression during development. Conversely, if the UAS downstream gene expresses a double-stranded interferent RNA (RNAi), one can lower the expression of a given gene in specific adult circuits (Fig. 3). Thus, the TARGET technique allows us to increase or decrease the expression of a precise gene in a temporal and regional manner.

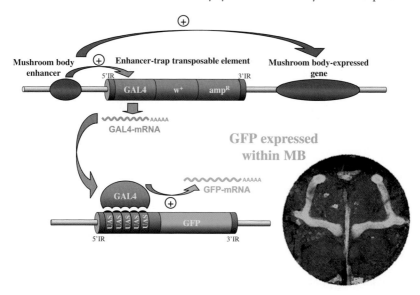

Fig. 2. The GAL4/UAS system. When an enhancer-trap transposable element is inserted near transcription enhancers that control expression in a given structure, the GAL4 gene that is contained in the P-element is expressed in the same structure. If this fly also contains a reporter gene downstream of the UAS sequences, GAL4 will transcribe the reporter. MB, mushroom body

		Development	**Adulthood**	**Phenotype**
A	GAL4, UAS-RNAi	MB-specific GAL4 / RNAi / RNAi / RNAi RNAi	MB-specific GAL4 / RNAi / RNAi / RNAi RNAi	**Memory defects possibly due to structural defects** (RNAi activity during development)
B	GAL4, UAS-RNAi GAL80ts	Low temperature MB-specific GAL4 GAL80ts / RNAi RNAi	High temperature GAL80ts MB-specific GAL4 / RNAi / RNAi / RNAi RNAi	**Specific Memory defects** (RNAi activity during adult life)

Fig. 3. The TARGET technique helps to discriminate a developmental gene action from an adult physiological effect. A sequence encoding an inverted repeat RNA (RNAi) is placed under UAS control. (**A**) GAL4 activates RNAi expression during development and adulthood. A memory defect could be caused indirectly by structural defects occurring during development. (**B**) The TARGET system solves this problem. Flies are maintained at low temperature during their developmental stages. At this temperature, GAL80^{ts} inhibits the transcription activity of GAL4, so no RNAi is expressed. When adults emerge, they are transferred at permissive temperature (30 °C); the GAL80^{ts} protein becomes ineffective and the GAL4 activates the transcription of the RNAi in adult MBs. The putative memory defects are due to the decreased concentration of the mRNA targeted by the RNAi

Gene-Switch uses a different mechanism to temporally control GAL4 activity (Burcin et al. 1999). The GAL4 DNA-binding domain was fused to a mutated progesterone receptor ligand-binding domain and part of the NF-B p65 activation domain. This synthetic transcription factor is active only when the antiprogestin, mifeprestone (RU486), is present. When flies are fed RU486, Gene-Switch is on and the UAS-transgene is expressed. As with the GAL4/UAS system, region-specific Gene-Switch activity is accomplished by the use of specific enhancer (Osterwalder et al. 2001; Roman et al. 2001). The use of these two techniques for olfactory memory studies will be detailed in the next section.

Finally, we recently developed a new GAL4 tool using an inducible chimerical protein that carries the Suppressor-of-hairy-wing repressor domain fused to the GAL4 DNA-binding domain. This tool also allows a temporally controlled repression of genes located near a P[UAS] insertion (Pascual et al. 2005).

Studying networks implicated in memory

Olfactory short-term memory

The MBs, centers of olfactory short-lasting memory

The insect brain contains a pair of prominent and characteristically shaped neural centers, the MBs. In *Drosophila*, MBs are composed of three main classes of neurons whose axons divide to form two vertical lobes (α and α') and three median ones (β, β'and γ) (Crittenden et al. 1998; Fig. 1). MBs are a specialized neuropile involved in processing and storing multimodal sensory information (Li and Strausfeld 1997). In the 1970s and '80s, the function of the MBs was assessed in different insect species with the classical interventionism approaches, cooling or ablation (Menzel et al. 1974; Erber et al. 1980). Over the years, MBs have been implicated in olfactory learning and memory and in a variety of complex functions, including courtship, motor control and spatial recognition.

Brain mutants with MB structural defects were isolated in the 1980s. The use of unlimited numbers of mutant animals with the same anatomical defect without interventionism helped to highlight the role of MBs in olfactory learning and memory (Heisenberg et al. 1985). However, one caveat of this approach is that the anatomical defects are not specific to the MBs (de Belle and Heisenberg 1994). An alternative approach was used to generate flies without MBs (de Belle and Heisenberg 1994). Hydroxyurea, an antimitotic agent, was fed to newly hatched wild-type larvae. At that early developmental stage, only five neuroblasts – the neurons' precursor cells – are mitotically active within each brain hemisphere: the four that generate the MBs and one in the antennal lobe. Thus, hydroxyurea treatment led to viable adult flies with almost no MBs. These flies turned out to show no olfactory memory (de Belle and Heisenberg 1994). The essential role of MBs in olfactory memory is further outlined by the high expression in the MBs of many of the proteins involved in learning and memory, such as Dnc (Nighorn et al. 1991), Rut (Han et al. 1992) and DCO (Skoulakis et al. 1993). Even though Amn (Quinn et al. 1979) is not expressed in the MBs, it is expressed by the dorsal paired median neurons (DPM) that project onto the MBs (Waddell et al. 2000).

Thus, *Drosophila* MBs are required for olfactory learning and memory, and insights gained from the first mutants have indicated that the cAMP pathway plays a key role for memory establishment. Nevertheless, it remained to be proven that cAMP metabolism was strictly required in MBs for correct learning and memory. Thanks to the GAL4/UAS system (Brand and Perrimon 1993), Connolly and colleagues (1996) disrupted MB cAMP signaling by expressing a constitutively activated G-protein ($G\alpha_s^*$). Permanent adenylyl cyclase activation led to an impaired associative memory.

That memory is impaired after ablation or functional disruption of a brain structure does not necessarily imply that the memory trace itself is localized within this structure. To localize short-term memory (STM), it was necessary to rescue the memory abilities of a mutant by expressing the corresponding protein in a specific brain structure. In a first step, Zars and colleagues (2000) expressed Rut Ca^{2+}/Calmodulin adenylyl cyclase in MBs to restore a normal learning capacity. However, in this experiment, Rut was expressed in the MBs at the adult stage but also during development. Thus, the behavioral rescue might have been due to the correction of a MB developmental defect, the indirect cause of the learning defect. Thanks to the TARGET and SWITCH methods (see above), McGuire and colleagues (2003) showed that the presence of Rut in adult MBs alone was sufficient to rescue *rut* memory defect. The current view is that Rut adenylyl cyclase could act during learning as a coincidence detector for the US and CS (McGuire et al., 2003).

Thanks to the GAL4/UAS system, Waddell et al. (2000) demonstrated a role for the Amn peptide in a well-defined group of cerebral neurons involved in middle-term memory formation. In the *amn* memory mutant, the *amn*[+] gene may be expressed in only two neurons (DPM) to re-establish normal olfactory memory (Waddell et al. 2000). DPM neurons project onto the MBs.

A tool to build anatomo-functional maps

A sophisticated approach to disturb neuronal circuits, based on a rapid and reversible blockage of synaptic transmission, was developed by Kitamoto (2001). The *shibire* (*shi*) gene encodes a microtubule-associated GTPase, Dynamin, which is involved in endocytosis and is essential for synaptic vesicle recycling and maintenance of the readily releasable pool of synaptic vesicles (Chen et al. 1991). The temperature-sensitive allele, *shi*[ts1], is defective in vesicle recycling at restrictive temperatures ($> 29\,°C$), resulting in a rapid blockage of synaptic transmission. The *shi*[ts1] mutation has a dominant effect because it blocks chemical synapses even in the presence of a normal *shi*[+] allele. The expression of *shi*[ts1] tool can therefore be used within the GAL4/UAS system. The GAL4/UAS-*shi*[ts1] approach is very powerful, as it allows the inhibition of particular brain circuits at a precise time. Are the MBs required during the acquisition, consolidation or retrieval phase? It was possible to address this question by expressing the temperature-sensitive Shi[ts1] protein specifically in MB neurons to transiently disrupt synaptic neurotransmission (Dubnau et al. 2001; McGuire et al. 2001). It was shown that the synaptic outputs of MBs neurons were required during retrieval of the STM but not during acquisition or consolidation. All together, these data indicate that a STM trace can be localized in MBs. Circuits that connect to the MBs have also been studied with Shi[ts1]. Thus, it has been established that abolishing vesicle-mediated secretion from the DPM in a precise time window, during the consolidation phase (Keene et al.

2004), phenocopies *amn* mutant memory defect (Waddell et al. 2000); likewise, blocking dopaminergic neurons (which project onto MB lobes) only during acquisition disrupts memory (Schwaerzel et al. 2003).

Imaging brain activity

One major caveat of drosophila central brain studies is that direct electrophysiological analysis is scarce (Wilson et al., 2004), due to the small size of neuron cell bodies (less than 5 μm in diameter). To circumvent this difficulty, two partially alternative approaches have been followed: analysis of learning memory mutants at the neuro-muscular junction (Zhong and Wu 1991; Renger et al. 2000), which leaves out the complexity of brain physiology, and analysis of isolated MB neurons in culture (Wright and Zhong 1995). Those experimental systems can provide interesting molecular and cellular information, but they are inadequate for assessing neuronal function at the level necessary for a global understanding of memory systems.

Odor processing occurs in a complex tissue environment, and identification of the repertoire of brain cell assemblies involved in olfactory memory requires visualization of the network activity at high spatial and temporal resolution, in preparations that are as intact as possible. Optical neural activity recordings allow study of active brains with micrometer-spatial resolution, and activity-sensitive fluorescent probes have been recently used in *Drosophila*. Interestingly, those sensors are proteins and therefore their expression can be driven to specific subsets of neurons with the GAL4/UAS system. Several sensors have been successfully brought to *Drosophila*; they monitor the local change of pH that accompanies neurotransmitter release (Yu et al. 2004) or changes in the intracellular calcium concentration that provide a valuable indicator of electrical activity [Aequorin (Rosay et al. 2001): Cameleon (Fiala et al. 2002), Camgaroo (Yu et al. 2003), and G-CaMP (Wang et al. 2003, 2004)]. Using the G-CaMP reporter and two-photon microscopy, stereotyped odor-evoked patterns have been observed in the antennal lobe glomeruli (Wang et al. 2003) and in the MBs (Wang et al. 2004).

The ultimate goal of imaging studies is to build a functional map of cell assemblies encoding memory in different regions of *Drosophila* brain by comparing the activity of trained and naïve animals, in normal flies or memory mutants. A first step was achieved when a transient change in the spatial code was observed in the antennal lobe of wild-type flies three min after olfactory associative conditioning (Yu et al. 2004). Another study has shown that, in naïve flies, electric shock generates a strong activity in dopaminergic neurons whereas the odor generates a weak signal. However, after several pairings between the odor and the shock, odor-evoked activity is significantly prolonged. In agreement with the behavioral approach (Schwaerzel et al. 2003), in vivo imaging therefore suggests that dopaminergic neurons play a role in aversive reinforcement in *Drosophila* (Riemensperger et al. 2005). In the same way, Yu et al. (2004) studied the function of the Amn-expressing DPM neurons in the memory process. DPM neurons responded to both the shock (the US) and to the odor (the CS), and pairing the CS and the US increased odor-evoked calcium signals in the same time window during which DPM neuron synaptic transmission was required for normal memory (Yu et al. 2004). The behavioral experiments using thermosensitive shibire (Waddell et al. 2000; Keene et al. 2004) were strengthened once again by imaging.

Olfactory LTM

As seen above, in addition to STM, *Drosophila* can display LTM after a spaced and repeated conditioning (long protocol; Tully et al. 1994). LTM can last several days and depends on de novo protein synthesis. Are MBs implicated in *Drosophila* LTM? To answer this question we analyzed the alpha-lobes absent (*ala*) mutant, which shows a peculiar MB phenotype. Ten percent of *ala* individuals possess all five MB lobes, 36% lack the horizontal β and β' lobes, and 4.5% lack the vertical α and α' lobes (the remaining sub-populations presented different MB phenotypes in the left and the right hemisphere; Pascual and Preat 2001). *ala* mutant flies were trained with the short or the long protocol, and we analyzed separately the brain of flies that had made the correct and the wrong choice during the memory test to calculate the memory score of each class of *ala* mutants. Flies lacking α/α' lobes displayed no LTM, although their short-lasting memories were wild-type. Flies with all lobes present or without β/β' lobes had a normal STM and LTM. Thus, MBs are necessary to perform LTM and, more particularly, the α/α' vertical lobes (Pascual and Preat 2001). It was further shown that, by expressing Shi^{ts1} in α/β lobes, α lobes outputs are required during LTM retrieval (Isabel et al., 2004).

If MBs play a pivotal role in *Drosophila* olfactory learning and memory, a recent study suggests that brain structures located outside the MBs also participate in olfactory memory (Pascual et al. 2004). We have recently shown that the *Drosophila* brain is asymmetric; a small structure expressing Fasciclin II, the asymmetric body, is present only in the right hemisphere. Interestingly, about 7% of Canton-Special wild-type flies present a bilateral structure. Those symmetric flies showed no LTM at four days, suggesting that asymmetry may be required for generating, maintaining or retrieving LTM (Pascual et al. 2004). The asymmetric body is located near the central complex, a structure that connects brain hemispheres. However, the functional links between the asymmetric body and other brain circuits remain to be sorted out.

In an elegant study, Ashraf et al. (2006) showed that protein synthesis at the synapse is required for LTM and that LTM formation depends on calcium/calmodulin-dependent Kinase II (CaMKII) signalling, a pathway also implicated in synaptic plasticity in mammals (Kelleher et al. 2004). By driving the expression of a tagged CAMKII in projection neurons that link olfactory sensory neurons to MBs, they proved that the recruitment of the CAMKII expression to post-synaptic sites in the antennal lobe glomeruli (Fig. 1) is required to induce LTM (Ashraf et al. 2006). By performing this expression in different mutant backgrounds, they showed that synaptic protein synthesis was regulated by the RNA interference silencing complex (RISC).

Operant visual learning and memory

In *Drosophila*, MBs seem especially involved in olfactory learning and memory and not in visual learning (Wolf et al. 1998). Flies can remember that a visual pattern represents a danger because this cue has been associated with heat (for review, see Heisenberg et al. 2001). Recently, using a variety of neurogenetic tools, Liu et al. (2006) identified a central brain structure, the fan-shaped body (FB), as being involved in visual memory (Fig. 4). After showing that the adenylate-cyclase Rut was required for the association between the visual pattern and the reinforcer, they restored a normal

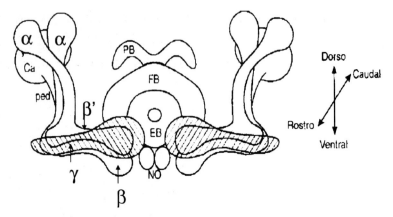

Fig. 4. Schematic representation of mushroom bodies (MB) and Central Complex (CX) neuropile. Dorsal is up; the most anterior is the first plane. MB: γ lobes, striped; α and β lobes; α' and β' lobes; ped, peduncle; Ca, calyx; CX: EB, ellipsoid body; FB, fan-shaped body; NO, nodulli; PB, protocerebral bridge

visual memory capacity in the *rut* mutant by specifically expressing the *rut* gene in the upper stratum of the FB (called the F5 neurons). Interestingly, F5 neurons are required to learn about the horizontal elevation of a cue, whereas the ability to learn about the contour orientation of the pattern involves FB F1 neurons (a lower horizontal stratum in FB). These results suggest that memories of two different visual features - elevation in the panorama and contour orientation - are stored in different groups of neurons within the same structure (Liu et al. 2006).

The dynamic of olfactory memory phases in *drosophila*

The short protocol induces two labile phases: STM, which is disrupted in mutants affected for cAMP metabolism and lasts about 30 min; and middle-term memory (MTM), which is disrupted in *amn* and lasts for a few hours. STM and MTM are anesthesia sensitive, as they are erased if flies are cooled down to 4 °C after conditioning. This property suggests that STM and MTM are sustained by electrical brain activities.

Drosophila also display two long-lasting memory phases, anesthesia-resistant memory (ARM; Quinn and Dudai 1976) and LTM (Tully et al. 1994; Isabel et al. 2004). ARM is generated by the short protocol or several massed trials. ARM can still be detected after a few days (Tully et al. 1994). This memory is disrupted in *rsh* mutants (Folkers et al. 2006). The atypical protein kinase M (aPKM) is a persistently active truncated isoform of atypical protein kinase C (aPKC). Overexpression of aPKM enhances memory after massed conditioning but not after spaced training (Drier et al. 2002). Moreover, inhibition of aPKM disrupts consolidated memory after massed conditioning (Drier et al. 2002). It is therefore conceivable that aPKC is a molecular support of ARM.

LTM is induced by the long protocol and measured for at least one week (Tully et al. 1994). Like ARM, LTM is anesthesia resistant (Isabel and Preat, unpublished results).

It is disrupted by a cAMP Response Binding Protein (dCREB2-b) repressor (Yin et al. 1994; Perazzona et al. 2004), likely due to the inhibition of gene expression required to establish LTM. Yin and colleagues (1995a,b) reported that flies overexpressing a particular CREB isoform (dCREB2-a) generated LTM after a single training cycle. However,

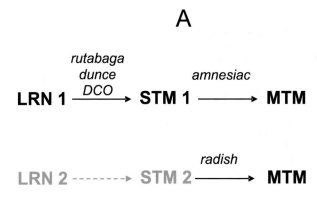

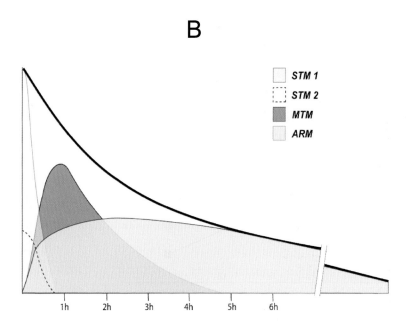

Fig. 5. Model of associative memory phases (**A**) and temporal dynamics of memory phases (**B**) generated by a single cycle of conditioning (short protocol). LRN: learning; STM: short-term memory; MTM: middle-term memory; ARM: anesthesia-resistant memory

this work could not be replicated, and it was shown that the original dCREB2-a trans-genic flies carried an accidental mutation that produced a truncated protein with no DNA binding domain (Perazzona et al. 2004). Moreover, adult induction of the correct CREB2-a isoform led to lethality (Perazzona et al. 2004).

What is the dynamic of memory phases in *Drosophila*? We proposed recently a model that involves two parallel memory pathways, one with cAMP dependent STM/MTM and the other with ARM (Fig. 5A). Indeed, *dnc* and *rut* retain a significant level of early memory (Tully and Quinn 1985), suggesting that an adenyl cyclase-Rut independent learning might exist. Moreover, ARM levels in *rut* and *amn* are near to normal (Folkers et al. 1993; Isabel et al. 2004), whereas their labile memories are strongly affected. Thus ARM does not seem to depend on STM/MTM. Instead a second learning process could give rise to a STM2 phases and later to ARM (Fig. 5).

What are the relationships between ARM and LTM? To answer that question, the *ala* mutant was trained with the long protocol and the memory of flies lacking vertical α/α' lobes was measured at 30 min and 5 hours after the training. Thirty-min memory was normal, but, surprisingly, five-hour memory was close to zero. Memory performance was normal at five hours when flies without vertical lobes were trained with the short protocol (Isabel et al. 2004; Fig. 6). Why does a longer training give rise to a weaker memory? *ala* flies display no LTM because they lack the vertical lobes, the center for LTM. These flies show a normal ARM five hours after the short protocol but they show no ARM after the long protocol. This result suggests that ARM is erased after LTM conditioning. Thus the consolidated memory phases generated by olfactory conditioning are exclusive (Fig. 7; Isabel et al., 2004). Why is ARM erased after LTM conditioning? We propose that ARM acts as a gating mechanism for LTM formation, avoiding a heavy cascade of gene expression in the absence of intensive

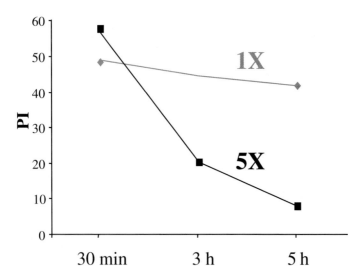

Fig. 6. In flies without MB alpha lobes, which normally sustain long-term memory, the long protocol decreases memory performance at five hours in comparison with the short protocol. *Gray line*, short protocol; *black line*, long protocol

A

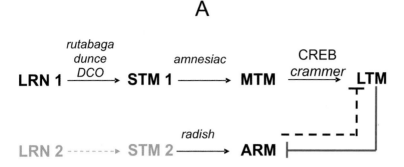

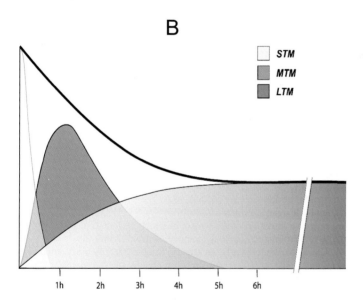

B

Fig. 7. Model of associative memory phases (**A**) and temporal dynamics of memory phases (**B**) generated by five cycles of conditioning (long protocol). LRN: learning; STM: short-term memory; MTM: middle-term memory; ARM: anesthesia-resistant memory; LTM: long-term memory

spaced conditioning. Despite the relative simplicity of the *Drosophila* brain, this model suggests a cognitive complexity more frequently associated with mammalian models. It supports the idea that *Drosophila* is a valid model to study some of the molecular and cellular mechanisms involved in normal or pathological human memory (Shulman et al. 2003).

References

Ashraf SI, McLoon AL, Sclarsic SM, Kunes S (2006) Synaptic protein synthesis associated with memory is regulated by the RISC pathway in Drosophila. Cell 124:191–205

Bellen HJ, O'Kane CJ, Wilson C, Grossniklaus U, Pearson RK, Gehring WJ (1989) P-element-mediated enhancer detection: a versatile method to study development in Drosophila. Genes Dev 3:1288–1300

Bier E (2005) Drosophila, the golden bug, emerges as a tool for human genetics. Nature Rev Genet 6:9–23

Boynton S, Tully T (1992) latheo, a new gene involved in associative learning and memory in Drosophila melanogaster, identified from P element mutagenesis. Genetics 131:655–672

Brand AH, Perrimon N (1993) Targeted gene expression as a means of altering cell fates and generating dominant phenotypes. Development 118:401–415

Burcin MM, Schiedner G, Kochanek S, Tsai SY, O'Malley BW (1999) Adenovirus-mediated regulable target gene expression in vivo. Proc Natl Acad Sci USA 96:355–360

Chen MS, Obar RA, Schroeder CC, Austin TW, Poodry CA, Wadsworth SC, Vallee RB (1991) Multiple forms of dynamin are encoded by shibire, a Drosophila gene involved in endocytosis. Nature 351:583–586

Chiang AS, Blum A, Barditch J, Chen YH, Chiu SL, Regulski M, Armstrong JD, Tully T, Dubnau J (2004) radish encodes a phospholipase-A2 and defines a neural circuit involved in anesthesia-resistant memory. Curr Biol 14:263–272

Comas D, Petit F, Preat T (2004) Drosophila long-term memory formation involves regulation of cathepsin activity. Nature 430:460–463

Connolly JB, Roberts IJ, Armstrong JD, Kaiser K, Forte M, Tully T, O'Kane CJ (1996) Associative learning disrupted by impaired Gs signaling in Drosophila mushroom bodies. Science 274:2104–2107

Crittenden JR, Skoulakis EM, Han KA, Kalderon D, Davis RL (1998) Tripartite mushroom body architecture revealed by antigenic markers. Learn Mem 5:38–51

Davis RL (2004) Olfactory learning. Neuron 44:31–48

Davis RL, Davidson N (1984) Isolation of the Drosophila melanogaster dunce chromosomal region and recombinational mapping of dunce sequences with restriction site polymorphisms as genetic markers. Mol Cell Biol 4:358–367

de Belle JS, Heisenberg M (1994) Associative odor learning in Drosophila abolished by chemical ablation of mushroom bodies. Science 263:692–695

DeZazzo J, Sandstrom D, de Belle S, Velinzon K, Smith P, Grady L, DelVecchio M, Ramaswami M, Tully T (2000) nalyot, a mutation of the Drosophila myb-related Adf1 transcription factor, disrupts synapse formation and olfactory memory. Neuron 27:145–158

Drier EA, Tello MK, Cowan M, Wu P, Blace N, Sacktor TC, Yin JC (2002) Memory enhancement and formation by atypical PKM activity in Drosophila melanogaster. Nature Neurosci 5:316–324

Dubnau J, Grady L, Kitamoto T, Tully T (2001) Disruption of neurotransmission in Drosophila mushroom body blocks retrieval but not acquisition of memory. Nature 411:476–480

Dubnau J, Chiang AS, Tully T (2003a) Neural substrates of memory: from synapse to system. J Neurobiol 54:238–253

Dubnau J, Chiang AS, Grady L, Barditch J, Gossweiler S, McNeil J, Smith P, Buldoc F, Scott R, Certa U, Broger C, Tully T (2003b) The staufen/pumilio pathway is involved in Drosophila long-term memory. Curr Biol 13:286–296

Erber J, Masuhr T, Menzel R (1980) Localization of short-term memory in the brain of the bee Apis mellifera. Physiol Entomol 5:343–358

Feany MB, Quinn WG (1995) A neuropeptide gene defined by the Drosophila memory mutant amnesiac. Science 268:869–873

Fiala A, Spall T, Diegelmann S, Eisermann B, Sachse S, Devaud JM, Buchner E, Galizia CG (2002) Genetically expressed cameleon in Drosophila melanogaster is used to visualize olfactory information in projection neurons. Curr Biol 12:1877–1884

Folkers E, Waddell S, Quinn WG (2006) The Drosophil radish Gene encodes a+protein required for anesthesia - resistant memory. Proc Natl Acad Sci 103:17496–17500

Folkers E, Drain P, Quinn WG (1993) Radish, a Drosophila mutant deficient in consolidated memory. Proc Natl Acad Sci USA 90:8123–8127

Grotewiel MS, Beck CD, Wu KH, Zhu XR, Davis RL (1998) Integrin-mediated short-term memory in Drosophila. Nature 391:455–460

Han PL, Levin LR, Reed RR, Davis RL (1992) Preferential expression of the Drosophila rutabaga gene in mushroom bodies, neural centers for learning in insects. Neuron 9:619–627

Heisenberg M (2003) Mushroom body memoir: from maps to models. Nature Rev Neurosci 4:266–275

Heisenberg M, Borst A, Wagner S, Byers D (1985) Drosophila mushroom body mutants are deficient in olfactory learning. J Neurogenet 2:1–30

Heisenberg M, Wolf R, Brembs B (2001) Flexibility in a single behavioral variable of Drosophila. Learn Mem 8:1–10

Isabel G, Pascual A, Preat T (2004) Exclusive consolidated memory phases in Drosophila. Science 304:1024–1027

Ito K, Suzuki K, Estes P, Ramaswami M, Yamamoto D, Strausfeld NJ (1998) The organization of extrinsic neurons and their implications in the functional roles of the mushroom bodies in Drosophila melanogaster Meigen. Learn Mem 5:52–77

Keene AC, Stratmann M, Keller A, Perrat PN, Vosshall LB, Waddell S (2004) Diverse odor-conditioned memories require uniquely timed dorsal paired medial neuron output. Neuron 44:521–533

Kelleher RJ, 3rd, Govindarajan A, Tonegawa S (2004) Translational regulatory mechanisms in persistent forms of synaptic plasticity. Neuron 44:59–73

Kitamoto T (2001) Conditional modification of behavior in Drosophila by targeted expression of a temperature-sensitive shibire allele in defined neurons. J Neurobiol 47:81–92

Leung B, Waddell S (2004) Four-dimensional gene expression control: memories on the fly. Trends Neurosci 27:511–513

Levin LR, Han PL, Hwang PM, Feinstein PG, Davis RL, Reed RR (1992) The Drosophila learning and memory gene rutabaga encodes a Ca^{2+}/Calmodulin-responsive adenylyl cyclase. Cell 68:479–489

Li Y, Strausfeld NJ (1997) Morphology and sensory modality of mushroom body extrinsic neurons in the brain of the cockroach, Periplaneta americana. J Comp Neurol 387:631–650

Liu G, Seiler H, Wen A, Zars T, Ito K, Wolf R, Heisenberg M, Liu L (2006) Distinct memory traces for two visual features in the Drosophila brain. Nature 439:551–556

Livingstone MS, Sziber PP, Quinn WG (1984) Loss of calcium/calmodulin responsiveness in adenylate cyclase of rutabaga, a Drosophila learning mutant. Cell 37:205–215

McGuire SE, Le PT, Davis RL (2001) The role of Drosophila mushroom body signaling in olfactory memory. Science 293:1330–1333

McGuire SE, Le PT, Osborn AJ, Matsumoto K, Davis RL (2003) Spatiotemporal rescue of memory dysfunction in Drosophila. Science 302:1765–1768

Menzel R, Erber J, Masuhr T (1974) Learning and memory in the honeybee. In: Barton-Brown L (ed) Experimental analysis of insect behaviour. Berlin, Germany: Springer, pp 195–217

Mohler JD (1977) Developmental genetics of the Drosophila egg. I. Identification of 59 sex-linked cistrons with maternal effects on embryonic development. Genetics 85:259–272

Moore MS, DeZazzo J, Luk AY, Tully T, Singh CM, Heberlein U (1998) Ethanol intoxication in Drosophila: Genetic and pharmacological evidence for regulation by the cAMP signaling pathway. Cell 93:997–1007

Nighorn A, Healy MJ, Davis RL (1991) The cyclic AMP phosphodiesterase encoded by the Drosophila dunce gene is concentrated in the mushroom body neuropil. Neuron 6:455–467

Osterwalder T, Yoon KS, White BH, Keshishian H (2001) A conditional tissue-specific transgene expression system using inducible GAL4. Proc Natl Acad Sci USA 98:12596–12601

Pascual A, Preat T (2001) Localization of long-term memory within the Drosophila mushroom body. Science 294:1115–1117

Pascual A, Huang KL, Neveu J, Preat T (2004) Neuroanatomy: brain asymmetry and long-term memory. Nature 427:605–606

Pascual A, Huang KL, Preat T (2005) Conditional UAS-targeted repression in Drosophila. Nucleic Acids Res 33:e7

Perazzona B, Isabel G, Preat T, Davis RL (2004) The role of cAMP response element-binding protein in Drosophila long-term memory. J Neurosci 24:8823–8828

Quinn WG, Dudai Y (1976) Memory phases in Drosophila. Nature 262:576–577

Quinn WG, Harris WA, Benzer S (1974) Conditioned behavior in Drosophila melanogaster. Proc Natl Acad Sci USA 71:708–712

Quinn WG, Sziber PP, Booker R (1979) The Drosophila memory mutant amnesiac. Nature 277:212–214

Renger JJ, Ueda A, Atwood HL, Govind CK, Wu CF (2000) Role of cAMP cascade in synaptic stability and plasticity: ultrastructural and physiological analyses of individual synaptic boutons in Drosophila memory mutants. J Neurosci 20:3980–3992.

Riemensperger T, Voller T, Stock P, Buchner E, Fiala A (2005) Punishment prediction by dopaminergic neurons in Drosophila. Curr Biol 15:1953–1960

Roman G, Endo K, Zong L, Davis RL (2001) P[Switch], a system for spatial and temporal control of gene expression in Drosophila melanogaster. Proc Natl Acad Sci USA 98:12602–12607

Rosay P, Armstrong JD, Wang Z, Kaiser K (2001) Synchronized neural activity in the Drosophila memory centers and its modulation by amnesiac. Neuron 30:759–770

Rubin GM, Yandell MD, Wortman JR, Gabor Miklos GL, Nelson CR, Hariharan IK, Fortini ME, Li PW, Apweiler R, Fleischmann W, Cherry JM, Henikoff S, Skupski MP, Misra S, Ashburner M, Birney E, Boguski MS, Brody T, Brokstein P, Celniker SE, Chervitz SA, Coates D, Cravchik A, Gabrielian A, Galle RF, Gelbart WM, George RA, Goldstein LS, Gong F, Guan P, Harris NL, Hay BA, Hoskins RA, Li J, Li Z, Hynes RO, Jones SJ, Kuehl PM, Lemaitre B, Littleton JT, Morrison DK, Mungall C, O'Farrell PH, Pickeral OK, Shue C, Vosshall LB, Zhang J, Zhao Q, Zheng XH, Lewis S (2000) Comparative genomics of the eukaryotes. Science 287:2204–2215

Schwaerzel M, Monastirioti M, Scholz H, Friggi-Grelin F, Birman S, Heisenberg M (2003) Dopamine and octopamine differentiate between aversive and appetitive olfactory memories in Drosophila. J Neurosci 23:10495–10502

Shimada T, Kato K, Kamikouchi A, Ito K (2005) Analysis of the distribution of the brain cells of the fruit fly by an automatic cell counting algorithm. Physica A 350: 144–149

Shulman JM, Shulman LM, Weiner WJ, Feany MB (2003) From fruit fly to bedside: translating lessons from Drosophila models of neurodegenerative disease. Curr Opin Neurol 16:443–449

Skoulakis EM, Davis RL (1996) Olfactory learning deficits in mutants for leonardo, a Drosophila gene encoding a 14-3-3 protein. Neuron 17:931–944

Skoulakis EM, Kalderon D, Davis RL (1993) Preferential expression in mushroom bodies of the catalytic subunit of protein kinase A and its role in learning and memory. Neuron 11:197–208

Strausfeld NJ, Sinakevitch I, Vilinsky I (2003) The mushroom bodies of Drosophila melanogaster: an immunocytological and golgi study of Kenyon cell organization in the calyces and lobes. Microsc Res Tech 62:151–169

Toba G, Ohsako T, Miyata N, Ohtsuka T, Seong KH, Aigaki T (1999) The gene search system. A method for efficient detection and rapid molecular identification of genes in Drosophila melanogaster. Genetics 151:725–737

Tully T, Quinn WG (1985) Classical conditioning and retention in normal and mutant Drosophila melanogaster. J Comp Physiol [A] 157:263–277

Tully T, Preat T, Boynton SC, Del Vecchio M (1994) Genetic dissection of consolidated memory in Drosophila. Cell 79:35–47

Waddell S, Quinn WG (2001) Flies, genes, and learning. Annu Rev Neurosci 24:1283–1309

Waddell S, Armstrong JD, Kitamoto T, Kaiser K, Quinn WG (2000) The amnesiac gene product is expressed in two neurons in the Drosophila brain that are critical for memory. Cell 103:805–813

Wang JW, Wong AM, Flores J, Vosshall LB, Axel R (2003) Two-photon calcium imaging reveals an odor-evoked map of activity in the fly brain. Cell 112:271–282

Wang Y, Guo HF, Pologruto TA, Hannan F, Hakker I, Svoboda K, Zhong Y (2004) Stereotyped odor-evoked activity in the mushroom body of Drosophila revealed by green fluorescent protein-based Ca^{2+} imaging. J Neurosci 24:6507–6514

Wilson C, Pearson RK, Bellen HJ, O'Kane CJ, Grossniklaus U, Gehring WJ (1989) P-element-mediated enhancer detection: an efficient method for isolating and characterizing developmentally regulated genes in Drosophila. Genes Dev 3:1301–1313

Wilson RI, Turner GC, Laurent G (2004) Transformation of olfactory representations in the Drosophila antennal lobe. Science 303:366–370

Wolf R, Wittig T, Liu L, Wustmann G, Eyding D, Heisenberg M (1998) Drosophila mushroom bodies are dispensable for visual, tactile, and motor learning. Learn Mem 5:166–178

Wright NJ, Zhong Y (1995) Characterization of K+ currents and the cAMP-dependent modulation in cultured Drosophila mushroom body neurons identified by lacZ expression. J Neurosci 15:1025–1034

Yang MY, Armstrong JD, Vilinsky I, Strausfeld NJ, Kaiser K (1995) Subdivision of the Drosophila mushroom bodies by enhancer-trap expression patterns. Neuron 15:45–54

Yin JC, Wallach JS, Del Vecchio M, Wilder EL, Zhou H, Quinn WG, Tully T (1994) Induction of a dominant negative CREB transgene specifically blocks long-term memory in Drosophila. Cell 79:49–58

Yin JC, Del Vecchio M, Zhou H, Tully T (1995a) CREB as a memory modulator: induced expression of a dCREB2 activator isoform enhances long-term memory in Drosophila. Cell 81:107–115

Yin JC, Wallach JS, Wilder EL, Klingensmith J, Dang D, Perrimon N, Zhou H, Tully T, Quinn WG (1995b) A Drosophila CREB/CREM homolog encodes multiple isoforms, including a cyclic AMP-dependent protein kinase-responsive transcriptional activator and antagonist. Mol Cell Biol 15:5123–5130

Yu D, Baird GS, Tsien RY, Davis RL (2003) Detection of calcium transients in Drosophila mushroom body neurons with camgaroo reporters. J Neurosci 23:64–72

Yu D, Ponomarev A, Davis RL (2004) Altered representation of the spatial code for odors after olfactory classical conditioning; memory trace formation by synaptic recruitment. Neuron 42:437–449

Zars T, Fischer M, Schulz R, Heisenberg M (2000) Localization of a short-term memory in Drosophila. Science 288:672–675

Zhong Y, Wu CF (1991) Altered synaptic plasticity in Drosophila memory mutants with a defective cyclic AMP cascade. Science 251:198–201

Towards a Molecular and Cellular Understanding of Remote Memory

Brian J. Wiltgen[1], Robert A.M. Brown[1], Lynn E. Talton[1], and Alcino J. Silva[1]

Summary

While the molecular, cellular and systems mechanisms required for the initial processing of memory have been intensively investigated by the new field of molecular and cellular cognition, those underlying permanent storage remain elusive. Here, we review neuroanatomical, pharmacological and genetic results demonstrating that specific areas of the cortex play a critical role in the storage of remote memory. Imaging experiments in rodents show that specific areas of the cortex are activated by remote memory and that this activation is impaired by manipulations that block remote memory. Accordingly, reversible inactivation of specific cortical structures in rodents disrupts remote memory without affecting recent memory. These results open a new, exciting window into memory studies and suggest a number of interesting experiments that will address molecular, cellular, systems and cognitive aspects of this poorly understood phase of memory.

Introduction

Memory is known to have several distinct phases, including a later phase in which memories are filtered and integrated into previous semantic constructs. This process is thought to involve molecular, cellular and systems processes that work in concert to transform and integrate information in the brain (McGaugh 2000; Debiec et al. 2002; Dudai 2004). In humans, research on the gradual transformation of memory representations over time has focused on declarative memories and their dependence on structures in the medial temporal lobe (MTL). These memories require the MTL and they (or some version of them) are thought to eventually be "stored" in neocortical circuits (Squire 1994; McClelland et al. 1995; Nadel and Moscovitch 1997). Results of studies of hippocampus-dependent memory in animals are largely consistent with this general idea (Zola-Morgan and Squire 1990; Kim and Fanselow 1992; Kim et al. 1993; Anagnostaras et al. 1999; Clark et al. 2002); but see also Sutherland et al. 2001). However, these studies have revealed remarkably little about the sites and mechanisms of remote aspects of memory. Similarly, the introduction of transgenic techniques has fueled an expansion of molecular and cellular studies of the very early stages (microseconds, seconds) of information processing in hippocampal networks (Mayford

---
[1] Departments of Neurobiology, Psychiatry and Biobehavioral Sciences, Psychology and Brain Research Institute. 695 Young Drive South, Room 2357 Box 951761, UCLA Los Angeles, California 90095-1761, USA
silvaa@mednet.ucla.edu

Bontempi et al.
Memories: Molecules and Circuits
© Springer-Verlag Berlin Heidelberg 2007

and Kandel 1999; Matynia et al. 2002; Tonegawa et al. 2003) but uncovered little about the mechanisms responsible for remote memory processing and integration in neocortical networks. A series of recent experiments, including studies from our own laboratory, demonstrate that specific regions of the neocortex and plastic mechanisms in these areas are critical for remote memory (Bontempi et al. 1999; Frankland et al. 2001, 2004; Takehara et al. 2003; Cui et al. 2004; Hayashi et al. 2004; Maviel et al. 2004; Remondes and Schuman 2004). One of these studies used an elegant genetic strategy to demonstrate that NMDAR function is critical for remote memory (Cui et al. 2004), and our laboratory showed that αCaMKII, one of NMDAR's key downstream effectors, is also involved in this memory phase (Frankland et al. 2001, 2004). Please note that a more elaborate description of many of the ideas and information included in this session was previously published in a review by the same authors (Wiltgen et al. 2004).

The involvement of multiple brain systems in remote memory

Determination of the specific systems involved in remote memory began with the finding that damage to the MTL produces severe amnesia. Patients with MTL damage have great difficulty forming new long-term memories (Scoville and Milner 1957; Penfield and Milner 1958; Corkin 1984). Additionally, MTL damage affects the acquisition of new declarative memories, while having a lesser effect on remote memories acquired long before the damage. These findings suggested that the MTL is essential for the initial acquisition and retrieval of declarative information but that eventually another structure, such as the neocortex, becomes involved in remote memory (Squire 1992). A related observation is that when brain pathology includes damage to the neocortex, remote memory is often impaired (Graham and Hodges 1997; Squire et al. 2001; Bayley et al. 2003). This observation suggests that neocortical areas are important for remote memory and that, although new memories are initially dependent on the MTL, they gradually become dependent on neocortical circuits (Alvarez and Squire 1994).

But why does the brain need two complementary memory systems? It is possible that gradual interleaving of memories in the neocortex is essential for the discovery of generalities and the eventual formation of knowledge structures (McClelland et al. 1995). The rapid and direct incorporation of new information into an intricate knowledge system such as the neocortex could cause catastrophic interference (McClelland et al. 1995). Instead, the hippocampus is thought to serve as a fast connector between distributed cortical memories, thus allowing these memories to be slowly integrated into existing knowledge systems in neocortical networks (McClelland et al. 1995). Besides its role in temporarily linking distributed cortical memories, the hippocampus is also thought to have a critical role in reactivating them. Reactivation by the hippocampus may serve to gradually strengthen the weak connections between memory fragments distributed in different neocortical sites (Buzsaki 1989) and may drive integration with previous related memories.

The neocortex and remote memory

The first evidence for neocortical engagement during in remote memory came from experiments in which animals were trained on a hippocampus-dependent spatial learning task and then monitored for brain activity [using (^{14}C)2-deoxyglucose uptake] following either recent (i.e., 5 days) or remote (> 25 days) memory tests (Bontempi et al. 1999). Retrieval of recent spatial memories produced more robust hippocampus activation than retrieval of remote memories, a result consistent with the model that proposes progressive disengagement of the hippocampus as memories are processed and integrated into semantic neocortical structures (Bontempi et al. 1999). In contrast, several neocortical areas studied, including the prefrontal cortex (PFC), showed the opposite pattern of activation, with more activation during remote memory tests. These results provided evidence in favor of the participation of cortical networks in remote memory processes. They revealed, for the first time, that specific neocortical areas do in fact become more engaged as memories become remote (Bontempi et al. 1999). These studies also pointed to specific regions of the neocortex that become activated in later stages of memory processing. The data revealed that, as memories change from recent to remote, the medial PFC [including the anterior cingulate cortex (ACC)], frontal, and temporal cortex all became more engaged (Bontempi et al. 1999). This finding allowed researchers to examine remote memory processes by targeting specific areas of the neocortex. Studies with this strategy, including our own, found that regions of the PFC are critical for the later stages of memory (Takehara et al. 2003; Frankland et al. 2004; Maviel et al. 2004). For example, hippocampal lesions in rats affect recent but not remote memory for trace conditioning (Takehara et al. 2003). In contrast, lesions of the medial PFC (including the ACC) produced the reverse gradient; they had only a small effect early (one day) after training but were devastating when made at later time points (two and four weeks). These results parallel prior activation data (Bontempi et al. 1999) and provide convergent evidence that activation of PFC regions, including the ACC, is critical for remote memory processes.

Our laboratory (Frankland et al. 2004) found increases in neocortical immediate early gene (IEG) expression following the retrieval of remote context fear memories in mice. IEGs can be used as markers of neuronal activation, and some of these genes, such as Zif268, ARC, etc., are also required for long-term potentiation and memory. We found that retrieval of recent memories produced robust IEG expression in the hippocampus but not in specific cortical areas such as the PFC and temporal cortex. However, retrieval of remote context memories produced the opposite effect, i.e.; educed IEG expression in the hippocampus and increased expression in temporal cortex and PFC areas, including ACC. The functional importance of this activation was demonstrated by blocking activity in the ACC during recent and remote memory tests: inactivation of the ACC impaired retrieval of remote (18 and 36 days post-training) but not recent (one and three days post-training) context fear memories. This finding suggests that, while the hippocampus is engaged, context fear memories do not require the PFC, but with the progressive disengagement of the hippocampus, the PFC becomes essential for late-stage memory processes. Similarly, recent findings showed increased IEG expression in neocortical regions (including ACC) following the retrieval of remote but not recent spatial memories in mice (Maviel et al. 2004). Targeted inactivation of these regions impaired the retrieval of remote but not recent spatial memory. Consistent with similar findings by our laboratory (Frankland et al. 2004), this study also revealed evidence that suggests remote memory triggers synaptic structural changes in the neocortex. Animals tested 30 days after training showed increases in the expression of the growth associated protein GAP43 compared to a 1-day retention group. This finding suggests that processing/integrating remote memories in cortical networks may involve synaptic structural changes. Another study suggesting that remote memory processes require synaptic remodeling involved mice with a dominant-negative transgenic PAK, a regulator of actin remodeling. The dominant-negative PAK transgene produced changes in plasticity and spine morphology in the cortex but not in the hippocampus. Accordingly, the mutant mice exhibited remote memory impairments (Hayashi et al. 2004). They exhibited normal recent but abnormal remote spatial memories. These results are consistent with findings indicating that experience-dependent changes in cortical function, such as those driven by sensory deprivation, are accompanied by synaptic structural changes (Trachtenberg et al. 2002).

In summary, specific areas of the neocortex are more activated by remote than by recent memory tests, and damage or inactivation of some of these activated regions selectively impairs remote but not recent memory. These studies also suggest that synaptic remodeling is an important facet of remote memory, suggesting that a dynamic reorganization of neocortical circuitry is required for integrating new information into pre-established semantic structures. If synaptic restructuring is occurring in specific neocortical areas during consolidation, one may ask what are the molecular and cellular mechanisms mediating these changes?

Molecular and cellular mechanisms of remote memory

Previous work showed that αCaMKII is essential for LTP and hippocampus-dependent learning (Matynia et al. 2002). Recently, our laboratory also found that this kinase plays a critical role in later stages of memory (Frankland et al. 2001). Spatial and contextual memory studies showed that a αCaMKII heterozygous null mutation (αCaMKII+/−) disrupted remote memory more severely than recent memory. Remarkably, electrophysiological analysis of these mice found normal hippocampal but impaired neocortical LTP (likely the result of robust αCaMKII expression in the hippocampus relative to the cortex). It is possible that the loss of cortical LTP interfered with some aspects of cortical remote memory and produced the unusual amnesic phenotype seen in these mice. The implication is that αCaMKII is a critical factor for the cortical plasticity underlying remote memory. Just as plasticity is thought to be required for the early stages of memory processing in hippocampal circuits, it may also be required for the progressive integration of this information into semantic cortical circuits. In a subsequent study (Frankland et al. 2001), our laboratory showed that the increases in neocortical activation, as measured by IEG induction, observed in normal mice following remote memory retrieval were completely absent in the αCaMKII mutants, as if plasticity or αCaMKII were a key requirement for the series of events that engage the cortex during remote memory processes. Hippocampal activation following recent memory retrieval and cortical activation immediately after training were normal in the mutants, whereas neocortical activation after remote memory was absent. This finding again suggests that remote memory depends critically on αCaMKII. Strikingly, elegant transgenic experiments demonstrated that a key activator of αCaMKII (NMDA receptor) is also critical for remote memory: an inducible and reversible NMDAR knockout was used to show that the retention of a 9-month fear memory is severely disrupted by extended, but not brief, loss of the NMDA receptor. Although this temporary disruption of NMDA receptors affected the prior remote memory, once completed it did not block the formation of new memories, suggesting that the blocking of remote memory was not caused by irreversible damage to cortical networks caused by the temporary loss of these receptors. Thus, it is possible that NMDA receptor-dependent activation of αCaMKII in neocortex is required for remote memory. NMDA receptors and αCaMKII are also known to be critical for both cortical LTP and for experience-dependent plasticity in the visual and somatosensory cortex (Fox 2002; Taha and Stryker 2005), suggesting that NMDAR-dependent activation of αCaMKII is a critical mechanism for cortical plasticity processes involved in both sensory plasticity and remote memory.

Mechanisms of structural plasticity: a possible role in remote memory

Previous studies demonstrate that αCaMKII messenger RNA is targeted to dendrites and that both the kinase and its activity are enhanced by synaptic activity (Soderling 2000). Elegant studies in the *Xenopus* retinotectal system demonstrated that this kinase has a direct modulatory role in the development of neuronal circuits (Cline 2001). These studies indicate that natural increases in αCaMKII expression during development cause a decrease in both new branch additions and retractions, thus stabilizing newly formed synapses (Wu et al. 1999; Cline 2001). But, is there any evidence that αCaMKII may also have a role in synaptic restructuring during cortical memory processes?

Remote memory retrieval leads to increases in the levels of GAP-43 in regions required for remote memory (Frankland et al. 2004; Maviel et al. 2004). GAP-43 is a synaptic protein with a role in synaptogenesis, synaptic plasticity and remodeling, as well as learning and memory (Benowitz and Routtenberg 1997). Thus, this increase in GAP-43 in specific neocortical areas during remote memory is consistent with the idea that remote memory processes are associated with synaptogenesis and synaptic remodeling (Wiltgen et al. 2004; Frankland and Bontempi 2005). Interestingly, the αCaMKII heterozygous mutants do not show the cortical increase in GAP-43 correlated with remote memory, an observation consistent with the hypothesis that remote memory may involve αCaMKII-dependent formation of new synaptic connections (Frankland et al. 2004). Findings in the somatosensory cortex are also consistent with the idea that αCaMKII modulates synaptic remodeling in cortical networks (Fox 2002).

Imaging cortical plasticity, imaging remote memory

Previous studies (Trachtenberg et al. 2002) used in vivo, time-lapse two- photon laser scanning microscopy (2P LSM) imaging to repeatedly study individual neurons in the barrel cortex of mice over many days and found that, although dendritic structure is stable, spine lifetimes vary greatly. A significant percentage of the spine population seems stable (lasting at least a month), whereas the rest appear to be far less stable (it is still unclear what the exact percentages are; Zuo et al. 2005a,b). Electron microscopic studies (Trachtenberg et al. 2002) indicated that synapse formation and elimination are associated with dynamic changes in spines. Importantly, experience-dependent changes in somatosensory function driven by whisker deprivation resulted in increased synapse turnover, a result consistent with the idea that storage of experience-dependent information drives changes in cortical spine structure (Trachtenberg et al. 2002; Zuo et al. 2005a,b). Physiological studies had demonstrated that αCaMKII has a key role in experience-dependent plasticity in the somatosensory cortex (Fox 2002). Together these studies are consistent with the idea that experience remodels cortical circuits and that this process is αCaMKII-dependent. These data also suggest that at least certain aspects of remote memory processing share mechanisms with sensory plasticity. In both cases the cortex is being asked to integrate previous representations with new ones. This hypothesis suggests several interesting questions: Are spines stable in regions required for remote memory? Does memory consolidation alter the stability of spines in cortical regions required for remote memory? Do manipulations that disrupt remote memory affect spine stability in cortical regions required for remote memory? These and a host of related questions suggest an exciting research program on remote memory that will involve state of the art transgenic, physiological, imaging, and behavioral approaches.

The immediate future of remote memory

The studies summarized above open the door to a systematic and comprehensive study of the molecular, cellular, systems and cognitive studies of remote memory. Additionally, there is a significant amount of conceptual and modeling work to be carried out that will affect the exciting upcoming experimental work. For example, despite a considerable amount of contradictory evidence, ideas about later stages of memory processing by and large assume a gradual processing and storage of information that remains essentially unchanged as it matures or consolidates in the brain (McGaugh 2000; Nadel et al., in press). Many of the conceptual frameworks used in memory storage have yet to integrate a growing body of data that has compellingly demonstrated that memory is not a process that faithfully records and stores information but an error-prone, dynamic and creative process that integrates new information with previous related semantic structures in a manner that is often more self-serving and subjective than faithful and reliable (Schacter and Norman 1998). Interestingly, our growing knowledge of the molecular and cellular mechanisms of memory, including remote memory, has unraveled a series of dynamic and probabilistic molecular and cellular mechanisms that often lack the faithful and reliable character that is so often assumed in studies of memory "consolidation". The very term "consolidation" implies a stability and permanence that memories rarely have. Indeed, recent studies have suggested that even the very process of recollection can change and alter supposedly "consolidated" memories (Nader et al. 2000; Sara 2000). In view of the surprising plasticity of acquired memories, one wonders whether the term "consolidation" has outlived its usefulness. Thus, it will be important to explore the character of remote memory with paradigms that more closely reflect the nature of remote memory. Although overtraining, strong emotional content, and simplicity can create memories that appear highly reliable and stable, real-life learning and ethologically meaningful memories rarely have these properties. Instead, more often than not memory has a tentative, imperfect and fuzzy character that is hardly captured in molecular, cellular and systems studies of memory, including remote memory. Thus, neuroscience studies should try to address the more interesting integrative and equivocal aspects of memory. Instead of failures of storage or recollection, they could potentially reflect highly adaptive and useful processes that remain unexplored.

How are remote memories stored in the brain? Are they distributed throughout the brain or are they stored close to the primary areas that initially processed them? Which regions "store" information, which regions organize retrieval, and more importantly, which circuits oversee the integration of new information into pre-established semantic structures? These are just a small sample of the vast array of cognitive neuroscience questions that need to be addressed concerning the neuroanatomy and processing of remote memory. In these studies, it will be important to not only image these processes but also to take full advantage of opportunities to study the impact of disruptions of these processes. Although often plagued with pleiotropic problems, human genetic disorders that affect cognitive function, from neurological to psychiatric conditions, offer precious opportunities to study brains with molecular and developmental dysfunction that may aid the cognitive neuroscience of memory, including remote memory. For example, rodent studies suggested a role for the prefrontal cortex in remote memory. It would be useful and informative to take advantage of the myriad of genetic conditions that disrupt pre-frontal function to evaluate the role of this structure in remote memory. It may not be sufficient to search for losses of remote memory among this varied patient population; instead it may be more productive instead to evaluate the nature, content and character of remote memories in these subjects.

Similarly, animal model studies in non-human primates, rodents, etc., also need to explore not only the neuroanatomy of remote memory but also the nature and behavioral characteristics of this form of memory (Nadel et al., in press). For example, recent studies suggest that in rodents, as in human subjects, remote memories lack the specificity and greater faithfulness of earlier memories (Wiltgen et al., 2006). It seems that, with time, memories become more general and less stimulus bound. These and other properties of remote memory need to be investigated and understood. It is clear that much needs to be done and that the studies summarized above have barely opened the door to this vast, but fascinating problem.

References

Alvarez P, Squire LR (1994) Memory consolidation and the medial temporal lobe: a simple network model. Proc Natl Acad Sci USA 91:7041–7045

Anagnostaras SG, Maren S, Fanselow MS (1999) Temporally graded retrograde amnesia of contextual fear after hippocampal damage in rats: within- subjects examination. J Neurosci 19:1106–1114

Bayley PJ, Hopkins RO, Squire LR (2003) Successful recollection of remote autobiographical memories by amnesic patients with medial temporal lobe lesions. Neuron 38:135–144

Benowitz LI, Routtenberg A (1997) GAP-43: an intrinsic determinant of neuronal development and plasticity. Trends Neurosci 20:84–91

Bontempi B, Laurent-Demir C, Destrade C, Jaffard R (1999) Time-dependent reorganization of brain circuitry underlying long-term memory storage. Nature 400:671–675

Buzsaki G (1989) Two-stage model of memory trace formation: A role for "noisy" brain states. Neuroscience 31:551–570

Clark RE, Broadbent NJ, Zola SM, Squire LR (2002) Anterograde amnesia and temporally graded retrograde amnesia for a nonspatial memory task after lesions of hippocampus and subiculum. J Neurosci 22:4663–4669

Cline HT (2001) Dendritic arbor development and synaptogenesis. Curr Opin Neurobiol 11:118–126

Corkin S (1984) Lasting consequences of bilateral medial temporal lobectomy: clinical course and experimental findings in H. M. Sem Neurol 4:249–259

Cui Z, Wang H, Tan Y, Zaia KA, Zhang S, Tsien JZ (2004) Inducible and reversible NR1 knockout reveals crucial role of the NMDA receptor in preserving remote memories in the brain. Neuron 41:781–793

Debiec J, LeDoux JE, Nader K (2002) Cellular and systems reconsolidation in the hippocampus. Neuron 36:527–538

Dudai Y (2004) The neurobiology of consolidations, or, how stable is the engram? Annu Rev Psychol 55:51–86

Fox K (2002) Anatomical pathways and molecular mechanisms for plasticity in the barrel cortex. Neuroscience 111:799–814

Frankland P, O'Brien C, Ohno M, Kirkwood A, Silva AJ (2001) Alpha-CaMKII-dependent plasticity in the cortex is required for permanent memory. Nature 411:309–313

Frankland PW, Bontempi B (2005) The organization of recent and remote memories. Nature Rev Neurosci 6:119–130

Frankland PW, Bontempi B, Talton LE, Kaczmarek L, Silva AJ (2004) The involvement of the anterior cingulate cortex in remote contextual fear memory. Science 304:881–883

Frankland PW, O'Brien C, Ohno M, Kirkwood A, Silva AJ (2001) alpha-CaMKII-dependent plasticity in the cortex is required for permanent memory. Nature (London) 411:309–313

Graham KS, Hodges JR (1997) Differentiating the roles of the hippocampal complex and the neocortex in long-term memory storage: evidence from the study of semantic dementia and Alzheimer's disease. Neuropsychology 11:77–89

Hayashi ML, Choi SY, Rao BS, Jung HY, Lee HK, Zhang D, Chattarji S, Kirkwood A, Tonegawa S (2004) Altered cortical synaptic morphology and impaired memory consolidation in forebrain-specific dominant-negative PAK transgenic mice. Neuron 42:773–787

Kim JJ, Fanselow MS (1992) Modality-specific retrograde amnesia of fear. Science 256:675–677

Kim KS, Lee MK, Carroll J, Joh TH (1993) Both the basal and inducible transcription of the tyrosine hydroxylase gene are dependent upon a cAMP response element. J Biol Chem 268:15689–15695

Matynia A, Kushner SA, Silva AJ (2002) Genetic approaches to molecular and cellular cognition: a focus on LTP and learning and memory. Annu Rev Genet 36:687–720

Maviel T, Durkin TP, Menzaghi F, Bontempi B (2004) Sites of neocortical reorganization critical for remote spatial memory. Science 305:96–99

Mayford M, Kandel ER (1999) Genetic approaches to memory storage. Trends Genet 15:463–470

McClelland JL, McNaughton BL, O'Reilly RC (1995) Why there are complementary learning systems in the hippocampus and neocortex: insights from the successes and failures of connectionist models of learning and memory. Psychol Rev 102:419–457

McGaugh JL (2000) Memory – a century of consolidation. Science 287:248–251

Nadel L, Moscovitch M (1997) Memory consolidation, retrograde amnesia and the hippocampal complex. Curr Opin Neurobiol 7:217–227

Nadel L, Winocur G, Ryan L & Moscovitch M (in press). Systems consolidation and the hippocampus: two views. Debates in Neuroscience.

Nader K, Schafe GE, LeDoux JE (2000) The labile nature of consolidation theory. Nature Rev Neurosci 1:216–219

Penfield W, Milner B (1958) Memory deficit produced by bilateral lesions in the hippocampal zone. AMA Arch Neurol Psych 79:475–497

Remondes M, Schuman EM (2004) Role for a cortical input to hippocampal area CA1 in the consolidation of a long-term memory. Nature 431:699–703

Sara SJ (2000) Retrieval and reconsolidation: toward a neurobiology of remembering. Learning Memory 7:73–84

Schacter DL, Norman KA (1998) The cognitive neuroscience of constructive memory. Annu Rev Psychol 49:289–318

Scoville WB, Milner B (1957) Loss of recent memory after hippocampal bilateral lesions. J Neurol Neurosurgical Psych 20:11–12

Soderling TR (2000) CaM-kinases: modulators of synaptic plasticity. Curr Opin Neurobiol 10:375–380

Squire LR (1992) Memory and the hippocampus; A synthesis from findings with rat, monkeys, and humans. Psychol Rev 99:195–231

Squire LR (1994) Memory and forgetting: long-term and gradual changes in memory storage. Int Rev Neurobiol 37:243–269; discussion 285–288

Squire LR, Clark RE, Knowlton BJ (2001) Retrograde amnesia. Hippocampus 11:50–55

Sutherland RJ, Weisend MP, Mumby D, Astur RS, Hanlon FM, Koerner A, Thomas MJ, Wu Y, Moses SN, Cole C, Hamilton DA, Hoesing JM (2001) Retrograde amnesia after hippocampal damage: recent vs. remote memories in two tasks. Hippocampus 11:27–42

Taha SA, Stryker MP (2005) Molecular substrates of plasticity in the developing visual cortex. Prog Brain Res 147:103–114

Takehara K, Kawahara S, Kirino Y (2003) Time-dependent reorganization of the brain components underlying memory retention in trace eyeblink conditioning. J Neurosci 23:9897–9905

Tonegawa S, Nakazawa K, Wilson MA (2003) Genetic neuroscience of mammalian learning and memory. Philos Trans R Soc Lond B Biol Sci 358:787–795

Trachtenberg JT, Chen BE, Knott GW, Feng G, Sanes JR, Welker E, Svoboda K (2002) Long-term in vivo imaging of experience-dependent synaptic plasticity in adult cortex. Nature 420:788–794

Wiltgen BJ, Brown RA, Talton LE, Silva AJ (2004) New circuits for old memories: the role of the neocortex in consolidation. Neuron 44:101–108

Wiltgen BJ, Matynia A & Silva AJ (2006). The dynamic nature of fear memories retained over long intervals. SFN online abstracts.

Wu GY, Zou DJ, Rajan I, Cline H (1999) Dendritic dynamics in vivo change during neuronal maturation. J Neurosci 19:4472–4483

Zola-Morgan SM, Squire LR (1990) The primate hippocampal formation: evidence for a time-limited role in memory storage. Science 250:288–290

Zuo Y, Lin A, Chang P, Gan WB (2005) Development of long-term dendritic spine stability in diverse regions of cerebral cortex. Neuron 46:181–189

Zuo Y, Yang GY, Kwon E, Gan WB (2005) Long-term sensory deprivation prevents dendritic spine loss in primary somatosensory cortex. Nature 436:261–265

Post-Activation State:
A Critical Rite of Passage of Memories

Yadin Dudai[1]

Summary

Each memory item has a unique biography. The global outline of the narrative of this biography, however, is shared by different items. The textbook account charts the ontogeny of memory items as a universal linear process. First the new information is encoded. Then it enters a short-term persistence phase, during which it is prone to interference by various types of treatments, ranging from distracting sensory stimuli to physical agents and drugs, collectively termed "amnesic agents". The period during which the item becomes gradually immune to the effect of amnesic agents is termed "memory consolidation". Upon completion of consolidation, so goes the zeitgeist, the stabilized item enters into a long-term "store", from which it can be later retrieved for use.

The textbook account may well be wrong. First and foremost, many lines of evidence suggest that "storage" is a misguided metaphor and that items in long-term memory are not stored but rather reconstructed each time they are retrieved (Bartlett 1932; Tulving 1983; Schacter 2001). Furthermore, this reconstruction might generate an internal representation that fails to faithfully replicate the representation of the original event (Bartlett 1932; Loftus and Loftus 1980; Schacter 2001). But even more relevant to the present discussion is the assumption that memory consolidates just once per item. Again and again, data have been reported that could be interpreted to indicate that this is not the case and that long-term memory items "reconsolidate" upon their reactivation. The latter hypothesis has been revitalized in recent years (Nader et al. 2000; Sara 2000; Nader 2003; Dudai 2004). The term "reconsolidation" itself is probably a misnomer, since whatever process the trace undergoes after its reactivation, faithful recapitulation of the original consolidation it is not. But few will argue with the data: something special and interesting happens to long-term memories following their use (Dudai 2006). Understanding what this "interesting thing" is, is highly relevant to our understanding of the nature of memory in general. Furthermore, what happens is not only of interest to armchair contemplation of memory: it may have important clinical implications, such as the ability to ameliorate post-traumatic memories (e.g., Miller et al. 2004).

The post-retrieval fate of items in long-term memory is the focus of this chapter. First I will briefly present data from an experimental system that has been studied in our laboratory for several years now and offers advantages for the study of consolidation as well as of post-retrieval processes. This system is conditioned taste aversion (CTA)

---

[1] Department of Neurobiology, The Weizmann Institute of Science, Rehovot 76100, Israel
yadin.dudai@weizmann.ac.il

Bontempi et al.
Memories: Molecules and Circuits
© Springer-Verlag Berlin Heidelberg 2007

in the rat. I will then proceed to propose that to properly construe the data on the formation, stability and post-retrieval fate of long-term memory items, we might wish to postulate the existence of a special memory state, which I will call the post-activation state (PAS). This state, so goes the claim, plays a critical role in the life of memories.

The fates of the CTA trace

CTA is the learned association of taste with visceral distress (Garcia et al. 1955; Bures et al. 1998; Rosenblum et al. 1993). In the typical protocol employed in our laboratory (Rosenblum et al. 1993), the rat is presented with a pipette containing a solution of an unfamiliar tastant, such as saccharin or glycine and, an hour or so later, is injected i.p. with a solution of LiCl, which produces visceral malaise for about two hours. In the test, usually performed a few days later, the rat is presented with a choice of two pipettes, one containing the malaise-associated tastant, the other a familiar, non-conditioned tastant. Conditioned rats prefer the non-conditioned over the conditioned tastant. In terms of classical (Pavlovian, elemental) conditioning, one might consider the conditioned tastant to be the conditioned stimulus (CS), the malaise-inducing agent to be the unconditioned stimulus (UCS), and taste rejection to be the conditioned response (CR). Many tastants can be used as CS, and other aversive agents, such as anxiogenic drugs, can be substituted for the LiCl as the UCS (Guitton and Dudai 2004). In contrast with other protocols of classical conditioning, which do not tolerate a CS-UCS interstimulus interval (ISI) of longer than a few seconds, CTA can tolerate an ISI of up to a few hours, implying that one can use the paradigm to investigate incidental learning (that of the CS in isolation, i.e. the formation of a novel taste memory) in addition to associative learning.

CTA is readily obtained after a single training trial, and the resulting memory lasts for many months (Berman et al. 2003) if not for a lifetime. However, if tested repeatedly in the absence of the UCS, memory extinguishes rather readily (Berman and Dudai 2001). CTA memory can, nevertheless, be rendered quite resistant to extinction if two training trials, 24 hours apart, are used instead of a single training trial (Berman et al. 2003; Eisenberg et al. 2003). Thus CTA is fit to be used in the investigation of fast incidental and associative encoding, long-term memory, and extinction.

Since CTA is readily induced in the laboratory rat, it can be analyzed at multiple levels, ranging from the behavioral to the molecular, or vice versa. In our laboratory, we use a combination of methods that span such multiple levels of analysis, ranging from behavioral studies, via lesions (both ablations and cytotoxic), cellular recordings in the behaving rat, targeted pharmacology (by microinfusions of selected agents into identified brain areas), and biochemistry and molecular biology of selected brain areas (Rosenblum et al. 1993, 1995, 1996, 1997; Lamprecht and Dudai 1995, 1996; Naor and Dudai 1996; Lamprecht et al. 1997; Berman et al. 1998, 2000; Berman and Dudai 2001; Bahar et al. 2003, 2004a,b; Desmedt et al. 2003). Most of our studies so far have focused on two major stations in the central pathways that subserve CTA: the taste cortex in the insular cortex (abbreviated IC), and two amygdalar nuclei, the central (CeA) and the basolateral (BLA) amygdala (Fig. 1).

All in all, our data support a flowchart model for the formation of the novel taste memory in the context of the CTA paradigm, with the following major attributes (Fig. 2):

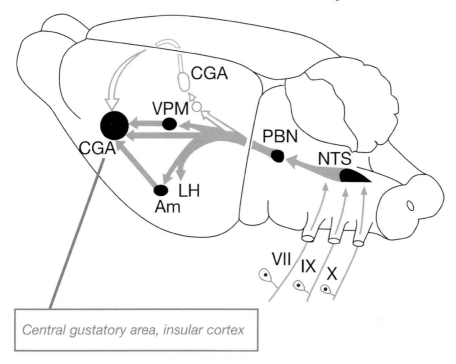

Central gustatory area, insular cortex

Fig. 1. The central taste system in the rat, which subserves taste learning and memory. Information from the buccal cavity flows via the cranial nerves (VII, IX and X) to the nucleus of the solitary tract (NTS), the parabrachial nucleus (PBN), and then via multiple pathways, either directly, or via the thalamus (ventroposteriomedial nuclei, VPM) or the amygdala (Am), to the central gustatory area (CGA) in the insular cortex (IC). LH, lateral hypothalamus

on-line information about a tastant converges on the IC. The incoming stimulus also activates neuromodulatory systems, specifically the cholinergic and the noradrenergic, that feed onto the IC. The incoming taste information (i.e. the CS information) is compared in the brain with off-line taste information, stored in the IC as well. The comparator circuit, which functions on a sub-second level, has not yet been identified, though it is likely to reside in the IC itself. If a match is detected, i.e., the tastant is deemed as familiar and the comparator circuit abates the activation of the aforementioned neuromodulatory systems. In the presence of a mismatch, i.e., unfamiliar taste, the neuromodulatory systems proceed to trigger a set of intracellular cascades in the IC, involving, among others, mitogen-activated protein kinases (MAPK), cAMP-responsive binding elements proteins (CREB) and their downstream substrates such as Elk1. This process culminates in modulation of gene expression and long-lasting synaptic and circuit modifications, which are assumed to embody long-term memory. The default reaction of the brain to a tastant is thus to consider that tastant unfamiliar and encode its attributes. This makes sense: unfamiliar tastes may kill, and false negatives are much more risky than false positives. Two additional points are noteworthy concerning this model. First, it doesn't specify where the CS-US association takes place. The association sites in CTA are still a matter of debate (e.g. Bures et al. 1998). There is ample evidence, however, that the IC contributes to the formation of the association, and, furthermore,

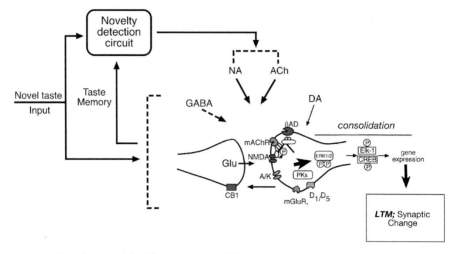

Fig. 2. A flowchart model of the acquisition of long-term taste memory. On-line information about a tastant converges on the IC. The stimulus also activates the cholinergic (ACh) and the noradrenergic (NA) neuromodulatory systems, which feed onto IC neurons. The incoming information is compared with off-line taste information, stored in the IC as well (novelty detection circuit). If a match is detected, the comparator abates the activation of the neuromodulatory systems. In the presence of a mismatch, i.e., unfamiliar taste, these neuromodulatory systems proceed to trigger a set of intracellular cascades in the IC, involving, among others, mitogen-activated protein kinases (MAPK, ERK1/2), other protein kinases (PKs), cAMP-responsive binding elements proteins (CREB) and their downstream substrates such as Elk1, and culminating in modulation of gene expression and long-lasting synaptic and circuit modifications, which are assumed to subserve long-term memory. The default reaction of the brain to a tastant is to consider it unfamiliar and encode its attributes. Glu, glutamate; mAChR, muscarinic acetylcholine receptor; NMDA, N-methyl-D-aspartate glutamatergic receptor; A/K, AMPA/Kainate glutamatergic receptors; mGluR, metabotropic glutamatergic receptors; DA, dopamine; $D_{1,5}$, dopaminergic receptors; βAD, β-adrenergic, receptors PKs, protein kinare systems; CB1, cannabinoid receptor 1; LTM, long-term memory. The cellular location of the receptors in the scheme is merely for illustration

to its storage over time, once it has been formed (Stehberg & Dudai and Shema Dudai, in preparation). Second, given the critical importance of CTA for survival, it appears that evolution has embedded backup circuits in the brain so that some reduced level of CTA can be obtained even in the absence of IC; the IC, nevertheless, plays a major role in encoding and retaining CTA memory in the intact rat.

So much for the encoding of long-term taste memory. But what happens when the memory is retrieved? As noted above, upon retrieval in the absence of the reinforcer (i.e., the UCS), a single-trial encoded CTA memory starts to extinguish. Ample evidence, obtained from many different systems, indicates that extinction is relearning of a CS-noUCS association rather than unlearning of the CS-UCS association (Pavlov 1927; Rescorla 1996). This finding immediately indicates that, in the retrieval situation, at least two traces are playing a role: the original, CS-UCS association, and the CS-noUCS association. We have identified additional traces that play a role as well, for example, the trace of the retrieval experience itself: recalling something bad is itself a bad experience (Berman et al. 2003). In addition, shortly after retrieval, the original CS-UCS association may regain control over behavior, in a process aptly termed "spontaneous recovery". The post-retrieval period seems to be characterized by the expression of multiple traces that compete over the control of the organism's behavior (Berman et al. 2003; Eisenberg et al. 2003; Fig. 3).

Memory consolidation windows are commonly delineated by testing the susceptibility of the memory trace to amnestic agents immediately after encoding. A widely used amnestic agent is the protein synthesis inhibitor, anisomycin. When microinfused into the IC within the first hour or so after training, but not afterwards, anisomycin blocks subsequent long-term CTA (Rosenblum et al. 1993). Short-term CTA is unaffected by this treatment. The question arises, since memory extinction is assumed to represent new learning rather than unlearning, will it also be blocked by microinfusion of anisomycin into the IC immediately after the extinction training (i.e., the retrieval trial in the absence of the reinforcer)? The answer is yes (Berman and Dudai 2001). Extinction, like the original association, consolidates.

But the extinction trace is not the only trace that can be blocked by the amnestic agent immediately after the retrieval session. As noted above, long-term CTA that is induced after two training trials (intensive training), 24 hrs apart, is rather resistant to extinction. Whereas application of anisomycin into the IC after retrieval of the regular training trace blocks extinction, the same treatment after retrieval of the intensive training trace results in amnesia of the original trace, rather than of the extinction trace. In other words, if the extinction trace controls behavior after retrieval, it becomes susceptible to the amnestic agent, whereas if the original trace controls behavior, it is this trace that becomes susceptible to the amnestic agent. Therefore, the susceptibility of the trace to the amnestic agent after retrieval is inversely proportional to trace dominance, where dominance is the ability of the trace to control behavior (Eisenberg et al. 2003; Fig. 4). This competition, observed at the behav-

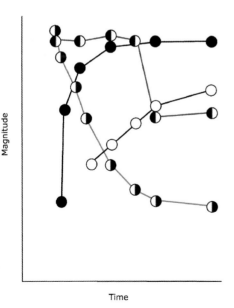

Fig. 3. Multiple memory traces become active following a non-reinforced retrieval of CTA memory. Among these traces are the expressed original CS-UCS association (excitatory trace, ◑), the new, CS-noUCS association (extinction, inhibitory trace,' ●) and the memory of the retrieval experience (◐). In addition, other associations of the retrieval cue may become active, and the original CS-UCS association spontaneously recovers over time in the absence of new taste input (○). The shape of the expressed traces in the figure is merely for the sake of illustration

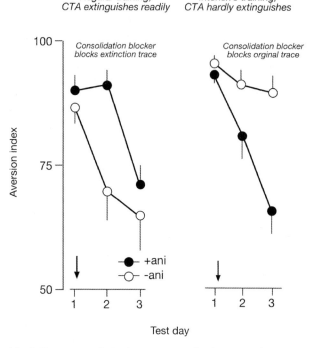

Fig. 4. Inverse correlation between trace dominance and post-retrieval sensitivity to an amnestic agent. The graphs show the effect of microinfusion of the protein synthesis inhibitor anisomycin, which is an amnestic agent, into the IC. Left panel: After a single CTA training trial (regular training), the trace extinguishes readily (–ani). Microinfusion of anisomycin into the insular cortex immediately after retrieval (*arrow*), under conditions that block consolidation of the original trace (Rosenblum et al. 1993), blocks extinction (+ani). Right panel: After two training trials, 1 day apart (intensive training), the trace becomes much more resistant to extinction (–ani). The effect of microinfusion of anisomycin into the insular cortex immediately after retrieval (arrow) is now reversed; the protein synthesis inhibitor leads to apparent amnesia of the original association (+ani). (Adapted from Eisenberg et al. 2003)

ioral level, may arise from competition of traces over shared plasticity resources at the cellular level. Competition over protein-synthesis-dependent plasticity resources has recently been demonstrated in a popular cellular model of plasticity, long-term potentiation (LTP), in the hippocampus (Fonseca et al. 2004). It is noteworthy that not all the post-retrieval traces are susceptible to the amnestic agent; the trace resulting from the CTA-retrieval experience is unaffected by anismoycin (Berman et al. 2003).

The post-retrieval CTA data are in line with the reconsolidation hypothesis, namely, that the reactivated long-term trace may have to undergo consolidation anew (Misanin et al. 1968; Sara 2000; Nader et al. 2000; Nader 2003; Dudai 2004). These data, however, do not prove that reconsolidation indeed takes place, let alone that it is recapitulation of consolidation. Ample evidence indicates that, in CTA, the molecular, circuit and behavioral signatures of the reconsolidation process differ from those of consolidation, and also from those of extinction (Bahar et al. 2004a; Kobilo, Stehberg and Dudai, unpublished). Differences between reconsolidation and consolidation have been demonstrated in other systems as well (reviewed in Dudai 2004, 2006). Indeed it would have been preferable to drop the term reconsolidation all together. However, this term has gained ground in the literature and is impractical to uproot – not unlike "memory extinction", which is not extinction at all.

The story is even more intriguing. Whereas consolidation has been detected in every species and long-term memory protocol analyzed so far, reconsolidation is not universal and appears to be constrained by boundary conditions, of which trace dominance (see above) is only one. In some systems, another boundary condition is the age of memory: older memories are rendered less amenable to reconsolidation (Suzuki et al. 2004; Eisenberg and Dudai 2004). Still another boundary condition is a mismatch between what the animal expects and what actually occurs; such a mismatch was reported to promote reconsolidation (Pedreira et al. 2004). Engagement of a memory-encoding mode in the reactivation session was also suggested to promote reconsolidation (Morris et al. 2006). It is noteworthy that the mismatch of the expected and the actual, i.e., surprise, drives encoding (Rescorla and Wagner 1972).

Post-Activation State (PAS)

The data on memory reconsolidation, of which the CTA data are only an example, are consistently evaluated in the literature within the context of the conceptual framework of the dual-trace hypothesis, which assumes the existence of two major memory phases (short and sensitive to amnestic agents, long and stable), and of consolidation, which converts the ephemeral trace into the seemingly secure world of engrams. My opinion is that this discussion is not very productive and reinforces mental stagnation, confined by the zeitgeist (for a similar attitude, see Nader et al. 2005). Paradigms deserve to be questioned occasionally. The current flare of data on the post-retrieval fate of the memory trace provides a proper opportunity to do just that. These data question the currently dominant dual-trace and consolidation (DTC) paradigm.

Provoked by the data and by earlier questioning of the dual-trace hypothesis (Lewis 1979) and of consolidation (Nadel and Moscovitch 1997), I propose recontemplation of the DTC paradigm. I do not consider reconsolidation to be recapitulation of a consolidation process; rather, it is a manifestation of a more general state in which the trace exists whenever it phases out of an activity period. What we consider reconsolidation is thus a temporarily altered state of the trace following memory reactivation. This post-activation state (PAS) is unveiled by increased sensitivity to amnestic agents, such as inhibitors of macromolecular synthesis (Dudai 2006). The increased sensitivity is probably a manifestation of enhanced plasticity of the neuronal circuit that encodes the memory trace or parts of it. Similarly, what we consider consolidation is a temporarily altered state of the trace following memory activation, again characterized by increased sensitivity to amnestic agents because of enhanced plasticity. The rule is, whenever the trace is active, it enters a state that is different from its inactive state. Reconsolidation and consolidation are different tokens of the same type of state, PAS. Seen this way, the trace doesn't exist in short- and long-term states but always alternates between active and inactive states (Lewis 1979). Active states occur in encoding (which in real-life involves retrieval of previous knowledge, as no brain is a tabula rasa), in retrieval (which in real-life involves encoding, as no retrieval replicates exactly the encoding state and context), and probably in background processing and maintenance of traces, such as in sleep (Siapas and Wilson 1998). PAS is a recurrent rite of passage of the engram, occurring whenever there is a transition between activity and inactivity of the circuit that encodes the internal representation. The circuit is never the same again.

What PAS is, what triggers it, and what is it for

What is the proposed PAS? The basic idea is that a surge of coherent activity in the neuronal circuit that encodes the internal representation relevant to the encoded, processed or retrieved item triggers a state of plasticity that is different from the period in which the representation is inactive. The fact that activity triggers plasticity is, of course, the keystone of the entire discipline of molecular neurobiology of learning, which rests on the assumption that teaching stimuli, encoded in action potentials or transmitters or neuromodulators, trigger synaptic and cell-wide plasticity mechanisms, which culminate in synaptic change (Kandel et al. 2000; Dudai 2002a). Here the expanded or added assumptions are: 1) activation of a representation induces a state of plasticity in critical nodes of the circuit that encodes that representation, regardless of whether the trigger for activation is external (i.e., sensory) or internal (i.e., other circuits in the brain); 2) this happens not only in encoding but also in retrieval and possibly also in implicit processing of the representation; 3) this state of plasticity outlasts the period of representational activity, yet terminates within minutes to a few hours at most; and 4) despite 1 and 2 above, differences are expected in the fine details of the induced plasticity because different states (e.g., encoded in global rhythms; Hasselmo et al. 2002; Csicsvari et al. 2003) allow the brain to distinguish between encoding, retrieval and processing. Since windows of enhanced plasticity disturb homeostasis (Dudai 2002a) and set in motion intracellular signal transduction cascades and translational control mechanisms, the system is likely to display modified sensitivity to stimuli and metabolic blockers. This transient modified sensitivity is assumed to provide the biochemical basis for the transient effectiveness of amnesic agents. Because different amnestic agents target different metabolic processes in the cell, one should expect the timing of the windows of sensitivity to amnestic agents to depend on the agent used, and this is indeed the case (Dudai 2004).

What is the biological function of PAS? First and foremost, one could question the question: assuming a role is the default of the adaptionist paradigm, which posits that biological phenomena always have phylogenetically selected roles (Gould and Lewontin 1979). It is still possible that PAS exists because the biological system cannot function otherwise, for example, because intracellular molecular networks may not be able to switch among (relatively) stable states instantaneously. But suppose the adaptionist point of view is valid. An appealing possibility is that PAS allows the system to better integrate the new information into the knowledge base of the organism, or update it. There is no lack of candidate cellular correlates for such integration, among them synaptic tagging (Frey and Morris 1997) and synaptic capture (Martin et al. 1997), all of which show that during the first minutes or hours after activation, neurons are particularly tuned to their recent history, integrate and redistribute inputs. In fact, the morphological re-modeling observed in some circuits following activation may represent the formation of associations and retrieval links, a must for efficient usability of information, rather than stabilization of the encoded new representation per se (Dudai 2004; Dudai and Eisenberg 2004). Another possibility is that PAS serves to strengthen traces that are used recurrently, because it is important that they remain functional over time (for an example of a cellular mechanism that might permit just that, see Tronson et al. 2006).

Seen this way, PAS is one of the three critical rites of passage of the engram. The first is encoding, in which the engram is born. The second is retrieval, in which the engram is expressed. The third is PAS, in which the engram is updated.

Some repercussions of the PAS hypothesis

On static vs. dynamic models of memory

Hebb, the most influential guide of modern dual-trace models, presented a dynamic universe of neuronal ensembles that encode internal representations (1949), but subsequent accounts of the Hebbian universe too often tended to depict the consolidated long-term trace as a relatively immutable mnemonic unit, falling for the lures of erroneous "storage metaphors" of memory persistence (discussed in Roediger 1980; Dudai 2004). The PAS hypothesis clearly argues for a dynamic view of long-term memory, in which internal representations are in continual flux; furthermore, it argues that the magnitude of this flux over time is proportional to the extent the representation is used. It is noteworthy that, in theory, flux doesn't necessarily imply significant change in content. Consider, for example, the classical Ship of Theseus problem, sometimes invoked in discussions of memory persistence (Dudai 2002a,b). The ship of the mythical Greek hero was placed on display in Athens, and with time, parts of it were replaced, one by one, until none of the original remained. Is this still the same ship? The same applies to other constructs, for example, the Pont Neuf in Paris, in which not a stone in the current structure remains in place from its original construction (Jones 2006). In both cases, however, flux retained the original structure. The PAS hypothesis, similarly to prominent cognitive accounts of memory (Loftus and Loftus 1980; Schacter 2001), assumes something more, namely, that the flux changes not only the physical substance but also the content and shape.

On metaphors

Some proponents of consolidation seem to account for the observed lingering sensitivity of declarative long-term memory items to amnesic agents by postulating that consolidation lingers for many years (Haist et al. 2001; for a different interpretation of this type of data, see Nadel and Moscovitch 1997). This extension of consolidation windows over decades questions the validity of the consolidation concept, as it raises the possibility that consolidation never ends, hence the trace is always prone to interference, hence consolidation does not exist. PAS, in contrast, does not annul the idea that some type of consolidation occurs; it only claims that, if consolidation does occur, it is relevant only until the subsequent activation of the relevant internal representation.

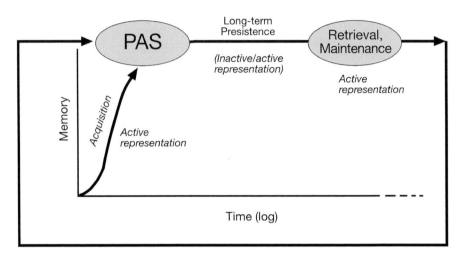

Fig. 5. Post-Activation State (PAS). The neuronal population that encodes the memory enters a special plastic state, characterized by increased sensitivity to amnestic agents each time it is active and phases into an inactive state. The inactive state is the state in which the internal representation encoded by the neuronal population is not coherently activated and is incapable of controlling behavior. The trace persists in an inactive state until further activation in retrieval or maintenance. Alternatively, the trace may also become transiently activated in maintenance, e.g. sleep. The expression of PAS at any given activation period of the trace is constrained by boundary conditions, such as competition with other active traces over plasticity resources or engagement of encoding mode in the activation session (see text). PAS comes in multiple mechanistic variants, and is manifested in the processes we call consolidation and reconsolidation. Memory traces are thus considered to exist in two alternating states, active and inactive (Lewis 1979), rather than in short- and long-term (consolidated) states. According to this view, each item in long term memory undergoes three types of rites of passage, encoding, retrieval, and PAS; items never really consolidate, or alternatively, consolidate only temporarily, from one PAS to another

In this respect, the PAS hypothesis can be said to convert the storage metaphor into a Phoenix metaphor: memories are reborn each time anew and mature only until their next reincarnation (Fig. 5). Metaphors in science should always be treated with caution, as they seldom depict reality; in the present case, whereas the mythical Phoenix dies only to be born anew, the mnemonic one doesn't die, only rejuvenates. Still, metaphors are not merely poetic devices that facilitate the communication of ideas. They are useful in guiding our scientific as much as our poetic thinking. Therefore, in the universe of metaphors, the Phoenix is more useful than the storage in guiding our research programs.

On working memory as a memory-updating space

Working memory is considered to be the memory system that holds information in temporary storage during the planning and execution of a task (Baddeley 2000; Repvos and Baddeley 2006). In doing so it is postulated to ephemerally combine on-line (sensory) and off-line (memory) information (Dudai 2004). The PAS hypothesis adds another dimension, or function, to working memory, which is providing updating space for long-term memory. Seen that way, working memory has two different, though clearly not mutually exclusive, roles: permitting the organism to combine present and past information for achieving an immediate goal, and enabling it to adjust knowledge about a changing world for future use.

On the nature of amnesia

Global amnesia has been portrayed over the years as a storage deficit (Squire 1987) and, by some, as a retrieval deficit (Warrington and Weiskrantz 1982; see also Riccio et al. 2003; de Hoz et al. 2004). It is possible that some types of amnesia are a PAS pathology. If PAS is truncated, or excessively sensitive to interference, amnesia will result. This is not merely semantics, as it may guide potential research avenues. Similarly, spontaneous confabulations may represent PAS deficits, in which updating retrieved information is faulty (Talland 1965).

On practical implications

Whereas potential prevention or amelioration of some types of amnesia is a remote possibility, prevention or amelioration of post-traumatic stress disorder (PTSD) is not. The revival of the reconsolidation literature (Sara 2000; Nader et al. 2000; Nader 2003) has incited attempts to modify or even erase long-term traumatic memories long after their formation. The PAS hypothesis supports such attempts. But additional practical implications of the PAS notion could be incited by assimilating into neuroscience a notion that is so familiar to cognitive psychologists, namely, that the life span of an item in memory persists only from one activation to another, and that memories, primarily declarative ones, change upon their use. Rites of passage, in which these changes are likely to take place, are not only windows of opportunity to peek into the mechanisms of memory; they also provide potential therapeutic windows.

Acknowledgements. I am grateful to Karim Nader, Joe LeDoux, Richard Morris, Mark Eisenberg, Jimmy Stehberg, and Tali Kobilo for valuable discussions; to the Volkswagen Foundation, the Human Frontiers Science Program (HFSP), the Israel Science Foundation (ISF), and the Nella and Leon Benoziyo Center for Neurological Diseases for support; and to the Fondation IPSEN, Paris, for organizing a marvelous, effective and educating meeting on *Memories: Molecules and Circuits*.

References

Baddeley A (2000) The episodic buffer: a new component of working memory? Trends Cogn Sci 4:417–423

Bahar A, Samuel A, Hazvi S, and Dudai Y (2003) The amygdalar circuit that acquires taste aversion memory differs from the circuit that extinguishes it. Eur J Neurosci 17:1527–1530

Bahar A, Dorfman N, Dudai Y (2004a) Amygdalar circuits required for either consolidation or extinction of taste aversion memory are not required for reconsolidation. Eur J Neurosci 19:1115–1118

Bahar A, Dudai Y, Ahissar E (2004b) Neuronal signature of familiarity in the taste cortex of the behaving rat. J Neurophysiol. 92:3298–3308

Bartlett FC (1932) Remembering. A study in experimental and social psychology. London: Cambridge University Press

Berman DE, Dudai, Y (2001) Memory extinction, learning anew, and learning the new: dissociations in the molecular machinery of learning in cortex. Science 291:2417–2419

Berman DE, Hazvi S, Rosenblum K, Seger R, Dudai Y (1998) Specific and differential activation of mitogen-activated protein kinase cascades by unfamiliar taste in the insular cortex of the behaving rat. J Neurosci 18:10037–10044

Berman DE, Hazvi S, Neduva V, Dudai Y (2000) The role of identified neurotransmitter systems in the response of the insular cortex to unfamiliar taste: activation of ERK1-2 and formation of a memory trace. J Neurosci 20:7017–7023

Berman DE, Hazvi S, Stehberg J, Bahar A, Dudai Y (2003) Conflicting processes in the extinction of conditioned taste aversion: behavioral and molecular aspects of latency, apparent stagnation, and spontaneous recovery. Learn Mem 10:16–25

Bures J, Bermudez-Rattoni F, Yamamoto T (1998) Conditioned taste aversion. Memory of a special kind. NY: Oxford University Press

Csicsvari J, Jamieson B, Wise KD, Buzsáki G (2003) Mechanisms of gamma oscillations in the hippocampus of the behaving rat. Neuron 37:311–322

de Hoz L, Martin SJ, Morris RGM (2004) Forgetting, reminding, and remembering: the retrieval of lost spatial memory. PLoS Biol 2:1233–1242

Desmedt A, Hazvi S, and Dudai Y (2003) Differential pattern of CREB activation in the rat brain after conditioned aversion as a function of the cognitive process engaged: Taste vs. context association. J Neurosci 23:6102–6110

Dudai Y (2002a) Memory from A to Z. Keywords, concepts and beyond. Oxford: Oxford University Press

Dudai Y (2002b) Molecular bases of long-term memories: a question of persistence. Curr Opin Neurobiol 12:211–216

Dudai Y (2004) The neurobiology of consolidations, or, how stable is the engram? Ann Rev Psychol 55:51–86

Dudai, Y (2006) Reconsolidation: the advantage of being refocused. Trends Cogn Sci 16:174–178

Dudai Y, Eisenberg M (2004) Rites of passage of the engram: reconsolidation and the lingering consolidation hypothesis. Neuron 44:93–100

Eisenberg M, Dudai Y (2004) Reconsolidation of fresh, remote, and extinguished fear memory in medaka: old fears don't die. Eur J Neurosci 20:3397–403

Eisenberg M, Kobilo T, Berman DE, Dudai Y (2003) Stability of retrieved memory: Inverse correlation with trace dominance. Science 301:1102–1104

Fonseca R, Nagerl UV, Morris RGM, Bonhoeffer T (2004) Competing for memory: hippocampal LTP under regimes of reduced protein synthesis. Neuron 44:1011–1020

Frey U, Morris RGM (1997) Synaptic tagging and long-term potentiation. Nature 385:533–536

Garcia J, Kimmeldorf DJ, Koelling RA (1955) Conditioned taste aversion to saccharin resulting from exposure to gamma radiation. Science 122:157–158

Gould SJ, Lewontin RC (1979) The spandrels of San Marco and the Panglosian paradigm: a critique of the adaptionist programme. Proc Roy Soc London B 205:581–598

Guitton MJ, Dudai Y (2004) Anxiety-like state associates with taste to produce conditioned taste aversion. Biol Psychiat 56:901–904

Haist F, Bowden Gore J, Mao H (2001) Consolidation of human memory over decades revealed by functional magnetic resonance imaging. Nature Neurosci 4:1139–1145

Hasselmo ME, Bodelon C, Wyble BP (2002) A proposed function for hippocampal theta rhythm: separate phases of encoding and retrieval enhance reversal of prior learning. Neural Comput 14:793–817

Hebb DO (1949) The organization of behavior: a neuropsychological theory. New York: Wiley.

Jones C (2006) Paris. A biography of a city. New York: Penguin Books

Kandel ER, Schwartz JH, Jessell TM (2000) Principles of neural science. New York: McGraw-Hill

Lamprecht R, Dudai Y (1995) Differential modulation of brain immediate early genes by intraperitoneal LiCl. NeuroReport 7:89–293

Lamprecht R, Dudai Y (1996) Transient expression of c-Fos in rat amygdala during training is required for encoding conditioned taste aversion memory. Learn Mem 3:31–41

Lamprecht R, Hazvi S, Dudai Y (1997) cAMP response element-binding protein in the amygdala is required for long- but not short-term conditioned taste aversion memory. J Neurosci 17:8443–8450

Lewis DJ (1979) Psychobiology of active and inactive memory. Psychol Bull 86:1054–1083

Loftus EF, Loftus GR (1980) On the permanence of stored information in the human brain. Am Psychol 35:409–420

Martin KC, Casadio A, Zhu H, Yaping E, Rose JC, Chen M, Bailey CH, Kandel ER (1997) Synapse-specific, long-term facilitation of Aplysia sensory to motor synapses: A function for local protein synthesis in memory storage. Cell 91:927–938

Miller MM, Altemus M, Debiec J, LeDoux JE, Phelps EA (2004) Propranolol impairs reconsolidation of conditioned fear in humans. Soc Neurosci Abst 208.2

Misanin JR, Miller RR, Lewis DJ (1968) Retrograde amnesia produced by electroconvulsive shock after reactivation of consolidated memory trace. Science 160:554–555

Morris RGM, Inglis J, Ainge JA, Olverman HJ, Tulloch J, Dudai Y, Kelly PAT (2006) Reconsolidation of spatial memory: Differential sensitivity of distinct spatial memory tasks to local inhibition of protein-synthesis in dorsal hippocampus following memory retrieval. Neuron 50:479–489

Nadel L, Moscovitch M (1997). Memory consolidation, retrograde amnesia and the hippocampal complex. Curr Opin Neurobiol 7:217–227

Nader K (2003) Memory traces unbound. Trends Neurosi 26:65–72

Nader K, Schafe GE, LeDoux JE (2000) Fear memories require protein synthesis in the amygdala for reconsolidation after retrieval. Nature 406:722–726

Nader K, Hardt O, Wang S-H (2005) Response to Alberini: right answer, wrong question. Trends Neurosci 28:346–347

Naor C, Dudai Y (1996) Transient impairment of cholinergic function in the rat insular cortex disrupts the encoding of taste in conditioned taste aversion. Behav Brain Res 79:61–67

Pavlov IP (1927) Conditioned reflexes. An investigation of the physiological activity of the cerebral cortex. London: Oxford University Press

Pedreira ME, Perez-Cuesta LM, Maldonado H (2004) Mismatch between what is expected and what actually occurs triggers memory reconsolidation or extinction. Learn Mem 11:579–585

Repvos G, Baddeley A (2006) The multi-component model of working memory: Explorations in experimental cognitive psychology. Neuroscience 139:5–21

Rescorla RA (1996) Preservation of Pavlovian associations through extinction. Q J Exp Psychol 49B:245–258

Rescorla RA, Wagner AR (1972) A theory of Pavlovian conditioning: variations in the effectiveness of reinforcement and nonreinforcement. In: Black AH, Proksay WF (eds) Classical conditioning II: Current research and theory. New York: Appleton-Century-Crofts, pp. 64–99

Riccio DC, Millin PM, Gisquet-Verrier P (2003) Retrograde amnesia: forgetting back. Curr Dir Psychol Sci 12:41–44

Roediger HI (1980) Memory metaphors in cognitive psychology. Mem Cogn 8:231–246

Rosenblum K, Meiri N, Dudai Y (1993) Taste memory: the role of protein synthesis in gustatory cortex. Behav Neural Biol 59:49–56

Rosenblum K, Schul R, Meiri N, Hadari Y, Zick Y, Dudai Y (1995) Modulation of protein tyrosine phosphorylation in rat insular cortex following conditioned taste aversion training. Proc Natl Acad Sci USA 92:1157–1161

Rosenblum K, Hazvi S, Berman DE, Dudai Y (1996) Carbachol mimics the effects of sensory input on tyrosine phosphorylation in cortex. NeuroReport 7:1401–1404

Rosenblum K, Berman DE, Hazvi S, Lamprecht R, Dudai Y (1997) NMDA receptor and the tyrosine phosphorylation of its 2B subunit in taste learning in the rat insular cortex. J Neurosci 17:5129–5135

Sara SJ (2000) Retrieval and reconsolidation: toward a neurobiology of remembering. Learn Mem 7:73–84

Schacter DL (2001) The seven sins of memory: how the mind forgets and remembers. Boston: Houghton Mifflin

Siapas AG, Wilson MA (1998) Coordinated interactions between hippocampal ripples and spindles during slow-wave sleep. Neuron 21:1123–1128

Squire LR (1987) Memory and brain. New York: Oxford University Press

Suzuki A, Josselyn SA, Frankland PW, Masushige S, Silva AJ, Kida S (2004) Memory reconsolidation and extinction have distinct temporal and biochemical signatures. J Neurosci 24:4787–4795

Talland GA (1965) Deranged memory. A psychonomic study of the amnesic syndrome. New York: Academic Press

Tronson NC, Wiseman SL, Olausson P, Taylor JR (2006) Bidirectional behavioral plasticity of memory reconsolidation depends on amygdalar protein kinase A. Nature Neurosci 9:167–169

Tulving E (1983) Elements of episodic memory. Oxford: Oxford University Press

Warrington EK, Weiskrantz L (1982) Amnesia: a disconnection syndrome? Neuropsychologia 20:233–248

Reactivation-Dependent Amnesia: Disrupting Memory Reconsolidation as a Novel Approach for the Treatment of Maladaptive Memory Disorders

Jonathan L.C. Lee[1] and *Barry J. Everitt*[1]

Summary

The persistence of memories reflects a dynamic, rather than a stable, process. The reactivation of a previously learned memory, for example by re-exposure to a conditioned stimulus, renders it labile, such that the concomitant administration of a variety of amnestic agents results in a reduced ability subsequently to retrieve that memory. This reactivation-dependent amnesia suggests that memories may undergo cycles of reactivation followed by a process of what has been called "reconsolidation". This review will discuss several of the central features of memory reconsolidation, taking a translational view of the potential of disrupting reconsolidation as a treatment strategy for maladaptive memory disorders such as post-traumatic stress and drug addiction.

It has long been observed that memories undergo a period of stabilization after their initial acquisition, during which time they are vulnerable to disruption (see Dudai, 2004 for review). The use of the term memory "consolidation" (or "konsolidierung" in the native German) to describe this phenomenon can be traced back to the writings of Müller and Pilzecker in 1900 (Lechner et al. 1999). Their studies of the learning and retrieval of verbal stimuli resulted in the suggestion that the "perseveration" or consolidation of stimuli may be necessary to strengthen the associations between them, and that interference with this perseveration process impairs associative learning. Importantly, delaying the interference resulted in a reduced disruption of subsequent retrieval.

It is notable that the phenomenon described by Müller and Pilzecker came to be explained within the framework of interference theory (Lechner et al. 1999). However, consolidation theory has persisted, with its core postulate that molecular events in the brain are set in train by a learning episode that takes time to stabilize the memory and ensure its persistence (McGaugh 2000). One important neuronal mechanism is the synthesis of new proteins that are hypothesized to enable the potentiation of existing synapses and the formation of new synapses (Bailey et al. 1996). Thus the blockade of protein synthesis during the post-learning period often resulted in subsequent amnesia (Davis and Squire 1984; Bourtchouladze et al. 1998; Schafe and LeDoux 2000). This retrograde amnesia exhibited a temporal gradient, similar to that observed by Müller and Pilzecker, such that delay of protein synthesis inhibition rendered it ineffective in

---
[1] Department of Experimental Psychology, MRC/Wellcome Trust Behavioural and Clinical Neuroscience Institute, University of Cambridge, Downing Street, Cambridge CB2 3EB, UK
jlcl2@cam.ac.uk; bje10@cam.ac.uk

Bontempi et al.
Memories: Molecules and Circuits
© Springer-Verlag Berlin Heidelberg 2007

causing amnesia. Therefore, central to consolidation theory is the assertion that there is a temporally limited window following learning during which the newly acquired memory can be disrupted by consolidation blockers, but after which the memory is stable and fixed in the brain (McGaugh 1966).

The expression of previously learned memories is known not to be constant over time. Several well-established factors, including context, motivation and extinction, influence the retrieval and behavioural expression of memories (Bouton and Moody 2004). However, in such accounts the persistence of the memory trace is not in question, consistent with the many observations that the application of consolidation blockers long after learning does not affect fully consolidated memories. Nevertheless, there is increasing evidence that the retrieval of memories results in them being once more vulnerable to disruption by consolidation blockers (so-called reactivation-dependent amnesia; see recent reviews by Nader 2003; Dudai and Eisenberg 2004; Alberini 2005). Early observations revealed that electroconvulsive shock, when applied many hours after fear learning, could result in amnesia for the fear memory, but only if rats were re-exposed to the training stimulus or outcome shortly before treatment (Misanin et al. 1968; Schneider and Sherman 1968). It was thus proposed that memories, instead of undergoing a single episode of consolidation, may cycle between active and inactive states, with the transition between these states being mediated by memory retrieval and reconsolidation (Lewis 1979; Nader 2003). It is a matter of current debate whether the disruption of memory reconsolidation reflects an impairment in the storage or retrieval of memories (Sara 2000; Millin et al. 2001; Nader 2003), in which the phenomena of spontaneous recovery, renewal and reinstatement are often discussed. However, regardless of the nature of the underlying impairment, blockade of memory reconsolidation has recently gained much interest as a potential therapeutic approach for neuropsychiatric disorders in which persistent maladaptive memories play a prominent role, such as post-traumatic stress disorder and drug addiction (Przybyslawski et al. 1999; Nader 2003; Lee et al. 2005; Miller and Marshall 2005; Valjent et al. 2006), and the present review will focus on this topic.

Given recent discussion regarding when the term "reconsolidation" should be used (Dudai 2006; Rudy et al. 2006), it is pertinent at this stage to stress that, for the purposes of this review, "reconsolidation" will be used to describe the putative process that is impaired when reactivation-dependent amnesia is observed subsequently. No assumptions are made concerning what the amnesia might reflect, either neurobiologically or psychologically, or what the purpose of memory reconsolidation might be. These questions are briefly considered at the end of the review.

Studies of aversive learning and memory reconsolidation

The recent resurgence in studies of reactivation-dependent amnesia results from the demonstration that this phenomenon could be observed following targeted intracerebral infusions of amnestic agents (Nader et al. 2000). This finding enabled the inference of a neurobiological mechanism underlying the amnestic effect, which had previously not been possible with the use of electroconvulsive shock and interference procedures. In particular, the use of the protein synthesis-inhibiting antibiotic, anisomycin, has become prevalent in the investigation of reactivation-dependent amnesia. Anisomycin has been infused directly into the basolateral amygdala (Nader et al. 2000; Duvarci and Nader 2004; Debiec et al. 2006), dorsal hippocampus (Debiec et al. 2002; Lee et al. 2004; Power et al. 2006) and insular cortex (Eisenberg et al. 2003) of rats to result in reactivation-dependent amnesia for discrete fear conditioning, contextual fear conditioning, inhibitory avoidance and conditioned taste aversion. Thus the apparent dependence of reactivated memories upon de novo protein synthesis has encouraged the view that reactivation-dependent amnesia is an important phenomenon providing an opportunity to investigate the mechanisms of memory persistence (Dudai 2006).

The protein synthesis-dependence of memory reconsolidation

Reconsolidation, like consolidation, may therefore involve the synthesis of new proteins that are necessary to restabilise reactivated memories (Nader et al. 2000; Nader 2003). However, some doubt must be expressed regarding the conclusions of studies that use anisomycin (Routtenberg and Rekart 2005; Rudy et al. 2006). In particular, anisomycin activates intracellular signaling cascades that themselves have been implicated in learning and memory, such as the mitogen-activated protein kinases (Bebien et al. 2003), and may also induce apoptosis (Iordanov et al. 1997). Therefore, the use of anisomycin alone is insufficient to establish the protein synthesis-dependence of memory consolidation and reconsolidation.

The involvement of transcription factors in memory consolidation and reconsolidation (Kida et al. 2002; Merlo et al. 2005) provides further evidence that protein synthesis is a functional requirement for these processes. The protein synthesis hypothesis, however, has been tested in a more specific manner, following the demonstrations that specific genes are upregulated following learning or retrieval episodes (Hall et al. 2000, 2001; Taubenfeld et al. 2001; Strekalova et al. 2003; von Hertzen and Giese 2005). The functional importance of the expression of these genes has been tested in rats, through the knockdown of their specific proteins by targeted infusions of antisense oligodeoxynucleotides (ASO) into the amygdala and hippocampus to impair memory consolidation (Guzowski and McGaugh 1997; Guzowski et al. 2000; Taubenfeld et al. 2001; Guzowski 2002; Lee et al. 2004; Malkani et al. 2004) and reconsolidation (Lee et al. 2004, 2005; Tronel et al. 2005). Given that ASO do not have the same non-specific and diverse effects as anisomycin, these studies provide further evidence that de novo protein synthesis in neurons of discrete brain loci, at least in the rat, is required for both consolidation and reconsolidation.

Reactivation-dependent amnesia for aversive memories has been observed across many other species (chicks, fish, snails, and crabs). However, many of these studies have involved general treatments such as anisomycin or anaesthesia (Litvin and Anokhin 2000; Anokhin et al. 2002; Pedreira et al. 2002, 2004; Eisenberg et al. 2003; Pedreira and Maldonado 2003; Eisenberg and Dudai 2004; Salinska et al. 2004; Sangha et al. 2004; Gainutdinova et al. 2005), and there are fewer reports of reconsolidation impairments with more specific treatments, such as inhibitors of NFκB (Merlo et al. 2005), glycosylation (Salinska et al. 2004), and antagonists of NMDA (Pedreira et al. 2002) and dopamine (Sherry et al. 2005) receptors. Thus it is important that studies across different species use a similar range of treatments to those employed in experiments on rodents to provide convergent evidence regarding the neurobiological basis of memory reconsolidation.

Reactivation-dependence of amnesia

Studies of reconsolidation have mostly included control groups in which the treatment, often anisomycin, is administered without re-exposing the subject to training stimuli to reactivate the memory. Such non-reactivation controls are essential to show the reactivation-dependent effects of amnestic agents (Nader 2003; Dudai 2004). Many studies generally either include treated and control non-reactivated groups, analyzing them independently to show that there is no significant difference between treatment and control in the non-reactivated condition (Nader et al. 2000; Debiec et al. 2002; Kida et al. 2002; Kelly et al. 2003; Lee et al. 2004; Suzuki et al. 2004; Merlo et al. 2005; Tronel et al. 2005), or they include only the treated non-reactivated group, showing that it is not different from control reactivated subjects but significantly different from those subjects both treated and reactivated (Przybyslawski et al. 1999; Anokhin et al. 2002; Milekic and Alberini 2002; Lattal and Abel 2004). However, to truly demonstrate the reactivation dependence of a treatment in comparison to control, it is necessary to show a significant Treatment x Reactivation interaction, where parametric statistics are permissible. Demonstration of a significant effect of reactivation in treated groups alone does not show that the effect of the amnestic agent depends upon the reactivation session, but only that the reactivation session itself has an effect on subsequent behaviour. Moreover, the independent analysis of reactivated and non-reactivated groups can be criticized for the adoption of repeated testing without adequate controls. Only a complete analysis of all groups can lead to the conclusion that the amnestic effect of a treatment is dependent upon the reactivation session. Relatively few studies have used such a rigorous design and analysis (Eisenberg et al. 2003; Sangha et al. 2003; Lee et al. 2005).

Therefore, to show the reactivation-dependent effects of amnestic agents unambiguously, the appropriate statistical analysis should be conducted wherever possible. On the other hand, observations of reactivation-*independent* amnesia need not necessarily be inconsistent with the phenomenon of reconsolidation. Little is known about the nature of memory reactivation, and it has been proposed that this might occur "implicitly as well as explicitly" (Alberini 2005) or "in the absence of behaviourally effective retrieval, e.g., in the course of background processing or sleep" (Dudai and Eisenberg 2004). Thus any reactivation-independent amnesia would be difficult to interpret and might in fact be expected in translational studies of aversive memory disorders such as post-traumatic stress, which is characterized in part by the subjective "re-experiencing" of the traumatic event (DSM-IV). It may instead be more important to pursue the investigation of the specificity of the amnesia, rather than its reactivation-dependence, as this will be an important factor to ensure that there are not widespread changes to memories as a result of treatment. A recent finding is reassuring in this regard, demonstrating that only directly reactivated memories undergo reconsolidation (Debiec et al. 2006).

Temporal gradients of reactivation-dependent retrograde amnesia

The temporal gradient of the disruptive effect of amnestic agents on memory reconsolidation following memory reactivation have also been investigated (Przybyslawski et al. 1999; Nader et al. 2000; Anokhin et al. 2002; Debiec et al. 2002; Pedreira et al. 2002;

Salinska et al. 2004). It was found consistently that there is a limited time window during which reactivated memories are vulnerable to disruption, and this temporal window may be shorter than that following initial learning (Debiec et al. 2002). However, again it could be argued that the independent analysis of delayed infusion groups, which often have smaller numbers of subjects, does not truly test for the temporal-dependence of the effects of amnestic agents, for which a combined analysis with an infusion time factor is more appropriate. Nevertheless, the consistent finding of a temporally graded amnestic effect strongly suggests that memories are transiently sensitive to amnestic agents following their reactivation.

It can be argued further that delayed infusion controls, though undoubtedly informative, are not necessary for the demonstration of reactivation-dependent amnesia and hence reconsolidation impairments. Rather, they help to define the parameters of memory reconsolidation. The temporal gradient of retrograde amnesia following memory reactivation parallels that following initial memory formation. However, the concept of a consolidation gradient is only useful within the theoretical framework of stable consolidated memories. The very existence of reactivation-dependent amnesia instead argues for a dynamic memory system in which memories cycle between active and inactive states (Lewis 1979; Nader 2003). Only if the time course of transformation between these memory states is the focus of a study should it be necessary or worthwhile investigating the temporal gradients of retrograde amnesia. This information may be especially useful in a translational setting to ensure that any treatment is administered within an effective time window following memory reactivation.

Another form of temporal gradient that is observed in studies of memory reconsolidation is that of a reduced amnestic effect when a greater period of time has elapsed between learning and reactivation-dependent treatment (Milekic and Alberini 2002; Eisenberg and Dudai 2004). However, old memories have been seen to be disrupted following reactivation (Debiec et al. 2002). It remains unclear whether memories generally become more resistant to reactivation-dependent amnesia with time, in which case it should always be possible to observe a temporally graded amnesia if sufficient time is allowed to elapse before memory reactivation. Alternatively, discrepant findings may be related to the different behavioural paradigms employed. The examples of a temporal gradient have been observed in inhibitory avoidance in rats (Milekic and Alberini 2002) and fear conditioning in *Medaka* fish (Eisenberg and Dudai 2004), but not with contextual fear conditioning (Debiec et al. 2002) or drug cue memory conditioning (Lee et al. 2006a) in rats. In either case, it will be important to define, in translational animal models, the temporal limitations of the use of reconsolidation blockade for the induction of amnesia for maladaptive memories.

Reconsolidation vs. consolidation

As highlighted above, commonalities between consolidation and reconsolidation in aversive memories include a similar dependence upon de novo protein synthesis and a limited temporal window of vulnerability. Experimental investigation of memory reconsolidation has frequently emphasized similarities between the two processes (Nader 2003; Nader et al. 2005), yet there are often notable differences in both the neuroanatomical and molecular domains (Lee et al. 2004; Alberini 2005). Thus reconsolidation may be vulnerable for a more limited period than consolidation (Judge and Quartermain 1982; Debiec et al. 2002), and it may also be more sensitive to disruption (Mactutus et al. 1979; Przybyslawski et al. 1999). Moreover, the neuroanatomical substrates of memory consolidation and reconsolidation appear to be dissociable, at least for some memories, such as those important for inhibitory avoidance and conditioned taste aversion (Taubenfeld et al. 2001; Bahar et al. 2004; Tronel et al. 2005). Finally, the molecular mechanisms of consolidation and reconsolidation are doubly dissociable (Lee et al. 2004), and reconsolidation is selectively impaired both by ASO for the immediate early genes *Zif268* (Lee et al., 2004) and C/EBPβ (Tronel et al. 2005), and by the beta adrenergic receptor antagonist propranolol (Debiec and LeDoux 2004). It can certainly be argued that any difference between consolidation and reconsolidation can be attributed to the inherent differences in experimental protocols, though this cannot explain examples of double dissociation (Lee et al. 2004; Nader et al. 2005; Dudai 2006).

Our previous finding that the molecular mechanisms in the hippocampus underlying contextual fear memory consolidation and reconsolidation are doubly dissociable (Lee et al. 2004) has, we suggest, several profound implications (Fig. 1). The effects of intra-hippocampal BDNF ASO to impair fear memory consolidation but not reconsolidation, and intra-hippocampal Zif268 ASO to disrupt reconsolidation only, cannot be explained as non-specific effects of ASO infusion into the hippocampus, for example in terms of general disruption of neuronal function or by state dependency. Rather, the effects are oligodeoxynucleotide sequence-dependent, strongly suggesting that they result from the knockdown of BDNF and Zif268 protein, respectively. This finding not only confirms that protein synthesis is required for both consolidation and reconsolidation but also demonstrates that the cellular mechanisms of these two processes are at

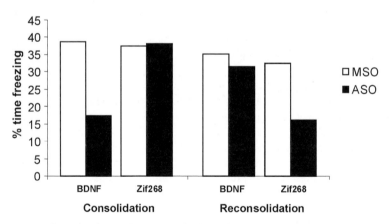

Fig. 1. The molecular mechanisms in the dorsal hippocampus of contextual fear memory consolidation and reconsolidation are doubly dissociable. Antisense oligodeoxynucleotides (ASO) for BDNF impaired consolidation, but not reconsolidation, when infused into the dorsal hippocampus 90 minutes before learning or reactivation. In contrast, intra-dorsal hippocampal Zif268 ASO selectively disrupted reconsolidation but not consolidation when infused at the same time points. Thus, the amnesia resulting from BDNF ASO and Zif268 ASO infusion cannot be explained by non-specific effects, and the cellular mechanisms of consolidation and reconsolidation are at least partially independent (Lee et al., 2004). MSO: control missense oligodeoxynucleotides

least partially independent. Therefore, it is not appropriate to describe reconsolidation in terms of recapitulating or partially recapitulating the consolidation process.

The utility of investigating and defining reconsolidation by constant reference and comparison to consolidation has recently been called into question (Dudai 2006). Commonalities and differences both in neurobiological mechanisms and psychological parameters do not provide conclusive evidence regarding the similarity or otherwise of the two processes. What seem like contentious issues have arisen more from the unfortunate terminology with which the field has been burdened. Consistent with our findings of a double dissociation in the underlying hippocampal molecular mechanisms of contextual fear memory consolidation and reconsolidation (Lee et al., 2004), it is generally accepted that reconsolidation is not a faithful recapitulation of consolidation at the mechanistic level (Alberini 2005; Nader et al. 2005; von Hertzen and Giese 2005; Dudai 2006). Therefore, the term "reconsolidation" should not be taken to imply mechanistic similarity, though the phenomenological commonalities remain undeniable (Nader et al. 2005).

Regardless of any theoretical value of comparing consolidation and reconsolidation at molecular, cellular or systems levels, there are certainly practical advantages to such an approach. Much is known about the neurobiological mechanisms that mediate the stabilization of memory following initial learning (Dudai 2004). The agreement that there is at least some overlap between consolidation and reconsolidation provides a useful starting point for the investigation of the molecular mechanisms of memory reconsolidation. However, we should no longer be surprised when consolidation and reconsolidation mechanisms are shown to be dissociable, either singly or doubly. Rather, one major challenge will be to delineate those mechanisms that are specific to memory reconsolidation (Debiec and LeDoux 2004; Lee et al. 2004; Tronel et al. 2005). This delineation is especially important from a translational viewpoint, as it will be most useful to disrupt selectively the reconsolidation of reactivated maladaptive memories, such as those that are important and debilitating in post-traumatic stress disorder, without the potential side effect of impairing new memories that are formed at the same time.

The discussion of reconsolidation as a phenomenon should also not be constrained by the longevity and dominance of consolidation theory. The very demonstration of reactivation-dependent amnesia is itself inconsistent with one interpretation of the consolidation hypothesis, namely that memories are fixed and immutable after an initial period of stabilization following learning. It should not be concluded, however, that retrieval-dependent amnesia invalidates the investigation of consolidation *as a phenomenon*; rather it questions some *theories* of consolidation. Consolidation and reconsolidation impairments resulting in amnesia are empirical findings that will enable a better understanding of the persistence and retrieval of memories (Dudai 2006), and future theories of memory will have to account for both phenomena. Indeed, modified theories of consolidation have been proposed that accommodate reactivation-dependent amnesia (Dudai and Eisenberg 2004; Alberini 2005). These theoretical positions have been criticised on the basis that they result in consolidation as a term losing heuristic value (Nader et al. 2005). However, this is a confusion that results precisely from the different meanings, theoretical and heuristic, of consolidation. There is in fact no dispute, as the word "consolidation" is being used in different ways. Perhaps it would be useful, when theorizing, to propose hypotheses of *memory* that encompass both *temporal gradients of retrograde amnesia* and *reactivation-dependent amnesia*, thereby obviating the use of the terms consolidation and reconsolidation entirely. Alternatively, the terms could be used exclusively for theoretical (Rudy et al. 2006) or empirical/heuristic meaning (Nader et al. 2005). However, the terminology is embedded as a useful explanatory tool in both circumstances, and it would be impractical, at least in the short to intermediate term, to change it in a coherent manner. Most importantly, we must strive as authors to be absolutely clear about the way in which we use the terms "consolidation" and "reconsolidation", and, as readers, not to make prior assumptions about how the terms are being employed.

Studies of non-aversive and appetitive learning and memory reconsolidation

The use of aversive learning paradigms in the study of learning and memory consolidation has been favored principally due to the utility of single session, or even single trial, learning. Similarly, investigation of memory reconsolidation has largely involved fear conditioning or avoidance memories. However, there is a growing literature on the reconsolidation of non-aversive and appetitive memories; in general, the findings are consistent with those of aversive conditioning studies.

Object recognition in rats is one example of rapidly acquired non-aversive learning, both the consolidation and reconsolidation of which appear to depend similarly upon protein synthesis in the ventromedial prefrontal cortex (Akirav and Maroun 2006) and ERK signaling in the hippocampus (Kelly et al. 2003). Zif268 knockout mice are also impaired both in the consolidation and the reconsolidation of object recognition (Bozon et al. 2003), although the latter of these findings will need to be verified, possibly with the use of inducible knockouts that do not also have initial learning/consolidation deficits. Reconsolidation impairments have also been observed in maze learning tasks in rodents (Przybyslawski and Sara 1997; Przybyslawski et al. 1999; Suzuki et al. 2004), as well as for odour-reward memories in both rats (Torras-Garcia et al. 2005) and honeybees (Stollhoff et al. 2005). Also of note are demonstrations of reactivation-dependent amnesia for incentive learning (Wang et al. 2005), eyelid conditioning (Inda et al. 2005), and motor learning (Walker et al. 2003), the last of which remains the only published demonstration to date of a reconsolidation impairment in humans.

Memory reconsolidation is thus not a phenomenon that is selectively restricted to aversive or rapidly acquired memories. However, it may not be universally applicable, as reactivation-dependent amnesia has yet to be observed for instrumental and contextual memories (Biedenkapp and Rudy 2004; Hernandez and Kelley 2004). Nevertheless, much interest has been taken in the demonstration of reactivation-dependent amnesia for addictive drug-related memories as a potential treatment strategy for drug addiction (Lee et al. 2005; Miller and Marshall 2005; Milekic et al. 2006; Valjent et al. 2006). Reconsolidation impairments have been observed in cocaine (Miller and Marshall 2005; Valjent et al. 2006) and morphine (Milekic et al. 2006) conditioned place preference (CPP) procedures, as well as in the control of cocaine seeking by cocaine cues (Lee et al. 2005); all of these studies were conducted in rodents. Importantly, the drug memory amnesia is both long-lasting (Lee et al. 2005; Miller and Marshall 2005; Milekic et al.

2006) and can be observed even after many pairings of a stimulus with *self-administered* cocaine (Lee et al. 2006a). It will be important to determine further the temporal dynamics of drug memory reconsolidation, especially with respect to the persistence of the amnesia, and whether older drug memories, established during protracted drug-taking histories that more closely model the characteristics of drug addiction in humans, are also susceptible to reactivation-dependent amnesia.

Memory reactivation

As highlighted previously, the nature of memory reactivation is an important focus of current research into reactivation-dependent amnesia. It has been suggested that there must be some mismatch between what occurs during memory reactivation and what is expected following initial training (Pedreira et al. 2004). Thus reactivation sessions in which both the stimulus and outcome are presented (so-called reinforced reactivation) preclude the observation of reactivation-dependent amnesia for aversive memories under some circumstances (Pedreira et al. 2004; Gainutdinova et al. 2005). On the other hand, several studies have observed reconsolidation impairments even when using reinforced reactivation (Sangha et al. 2003; Duvarci and Nader 2004; Eisenberg and Dudai 2004). However, it is unclear whether simply using an identical session to that of training, for example as is commonly the case in object recognition studies (Kelly et al. 2003; Akirav and Maroun 2006), is sufficient to eliminate any mismatch between expectancy and outcome, especially as such a "prediction error" is thought to be important for incremental learning (Rescorla and Wagner 1972; Schultz and Dickinson 2000). Thus, if the mismatch hypothesis is correct, reinforced reactivation should only be a boundary condition (Nader 2003; Dudai 2006) on the observation of reconsolidation impairments if subjects have previously been trained to asymptotic levels of performance such that further training trials generate no prediction error.

   A further factor that has been illustrated to great effect by the drug memory reconsolidation studies is that, sometimes, reinforced reactivation is *necessary* in order to provide conditions under which an impairment in reconsolidation can be seen following amnestic treatment. Miller and Marshall (2005) observed reactivation-dependent amnesia for a cocaine CPP when the reactivation session consisted of a preference test (rats were allowed to explore both cocaine-paired and unpaired compartments). However, concomitant administration of cocaine was apparently required when the rats were confined to the paired compartment during memory reactivation (Valjent et al. 2006). Furthermore, morphine administration in association with a preference test was necessary to induce a long-lasting amnesia for a morphine CPP (Milekic et al. 2006). These contrasting findings may be related to differences in the amnestic treatments employed (Valjent et al. 2006). Alternatively they may reflect the importance of the strength of reactivation, such that an increase in the strength of conditioning results in a greater degree of stimulus re-exposure being necessary to reactivate the memory (Suzuki et al. 2004). Exposure to the drug outcome may therefore be necessary to reactivate memories under certain circumstances, and the way in which memories are reactivated to enable their reconsolidation to be disrupted is a major consideration for both experimental and translational studies.

   Drug-conditioned place preference procedures have used three (Miller and Marshall 2005; Valjent et al. 2006) or 4 (Milekic et al. 2006) pairings of an environment with non-contingent injections of cocaine or morphine. This paradigm results in a significant preference for the paired compartment over a different compartment in which the subjects received injections of saline. However, CPP can be readily extinguished (Calcagnetti and Schechter 1993; Tzschentke and Schmidt 1995), unlike the effects of drug-associated cues to promote craving and relapse in addicts (Conklin and Tiffany 2002) and to support cocaine seeking in rats with a more extensive drug self-administration history (Di Ciano and Everitt 2004). Therefore, CPP may not model the strength and persistence of drug memories as effectively as drug self-administration procedures (Lee et al. 2005). It is notable that the studies of drug memory reconsolidation in rats self-administering cocaine have successfully demonstrated reactivation-dependent amnesia following re-exposure to a cocaine-associated cue alone, without the need for concomitant cocaine administration (Lee et al. 2005; Lee et al. 2006a). Therefore the apparent difficulty with which memories can be reactivated in the CPP model of drug addiction may not be inconsistent with our earlier statement that strong maladaptive memories might be especially vulnerable to amnestic treatment due to implicit or easily induced memory reactivation (see *reactivation-dependence of amnesia*). Indeed, exposure to cocaine-related videos or drug-related paraphernalia in abstinent cocaine addicts readily activates limbic cortical areas and induces craving for the drug (Grant et al. 1996; Childress et al. 1999; Garavan et al. 2000). It remains to be seen whether maladaptive memories in humans, both aversive and appetitive, are easy or difficult to reactivate such that their reconsolidation can be disrupted, and this issue will be intimately associated with that of the specificity of experimentally induced amnesia.

Translational applications of reactivation-dependent amnesia

Through the course of this review, several factors have been discussed that are of particular relevance to any translational application of reconsolidation blockade in the treatment of maladaptive memory disorders. These include empirical parameters such as whether old and well-established memories are vulnerable to reactivation-dependent amnesia, and the persistence of the amnesia so-induced. Specificity of the amnesia is also an important issue, as it would be problematic for any therapeutic strategy if closely related adaptive memories were also disrupted. The reactivation-dependence of reconsolidation impairments is a particularly advantageous feature, as it has the potential to limit to a high degree of specificity the effects of amnestic agents. However, the ways in which memories can be reactivated successfully are poorly understood and require further investigation.

The identification of tractable reconsolidation-specific neurobiological mechanisms will enable highly selective targeting of the maladaptive memories. Since the amnestic treatment in any future clinical application will have to be administered systemically, there should be a focus on identifying molecular mechanisms that are specific to memory reconsolidation. The observation that the beta-adrenergic receptor antagonist, propranolol, disrupts fear memory reconsolidation is therefore of great interest (Debiec and LeDoux 2004). However, careful consideration must be given to the effects of systemically administered drugs at several neuroanatomical loci and upon other memory mechanisms occurring in distributed neural systems. For example, propranolol can also impair emotional memory consolidation under certain circumstances (Przybyslawski et al. 1999; Cahill et al. 2000). Furthermore, whereas *Zif268* expression is required specifically in the hippocampus for the reconsolidation of contextual fear memories, it has also been implicated in the consolidation of contextual fear in the amygdala (Malkani et al. 2004), and Zif268 knockout mice are impaired in object recognition memory consolidation (Bozon et al. 2003). Similarly, the expression of *C/EBPβ* is specifically required for the consolidation of inhibitory avoidance memories in the hippocampus but also for their reconsolidation in the amygdala (Taubenfeld et al. 2001; Tronel et al. 2005). Therefore, as discrete neuroanatomical targeting of systemically administered pharmacological treatments is unlikely, the potential side effects of amnestic treatment must be thoroughly explored.

A final consideration relates to an alternative treatment strategy for maladaptive memory disorders, namely the potentiation of extinction (Davis 2002; Ressler et al. 2004). Extinction also depends upon de novo protein synthesis, and there is a fine balance between whether protein synthesis inhibitors will impair extinction to preserve memories or disrupt reconsolidation to result in amnesia (Eisenberg et al. 2003; Pedreira and Maldonado 2003; Suzuki et al. 2004). Similarly, both extinction and reconsolidation can be potentiated, either to diminish memories or to strengthen them (Walker et al. 2002; Ledgerwood et al. 2003; Frenkel et al. 2005; Tronson et al. 2006). Therefore, it will be of critical importance to ensure that attempts to disrupt the reconsolidation or enhance extinction do not instead result in the strengthening and increased persistence of maladaptive memories in humans (Lee et al. 2006b).

Concluding remarks

The observation that reactivation-dependent amnesia occurs consistently across different memory systems and species, using a variety of amnestic agents, strongly indicates that reconsolidation is a phenomenon that must be incorporated into theories of memory. This, however, does not address the fundamental questions of what reconsolidation is and what its function might be. It has been suggested that reconsolidation may enable memories to be restabilised (Nader 2003), updated (Dudai 2004; Dudai and Eisenberg 2004), or further stabilised and modified (Alberini 2005). Insights into such ideas may be provided either through novel behavioural approaches (de Hoz et al. 2004; Tronel et al. 2005) or by analysis of simpler animal models with better-defined neural networks (Sangha et al. 2003, 2004) or cellular analogues of learning and memory (Fonseca et al. 2006) to delineate the mechanisms of memory reactivation and reconsolidation at the synaptic level. Increasing our understanding of both the psychology and neurobiology of memory reconsolidation will, of course, enable a better characterization of the reactivation-dependent amnesia itself (including whether it can be best described as a storage or retrieval failure). However, regardless of the eventual answers to these questions, it is clear that the possibility of disrupting previously acquired maladaptive memories exists and warrants further study.

References

Akirav I, Maroun M (2006) Ventromedial Prefrontal cortex is obligatory for consolidation and reconsolidation of object recognition memory. Cereb Cortex 16:1759–1765

Alberini CM (2005) Mechanisms of memory stabilization: are consolidation and reconsolidation similar or distinct processes? Trends Neurosci 28:51–56

Anokhin KV, Tiunova AA, Rose SP (2002) Reminder effects - reconsolidation or retrieval deficit? Pharmacological dissection with protein synthesis inhibitors following reminder for a passive-avoidance task in young chicks. Eur J Neurosci 15:1759–1765

Bahar A, Dorfman N, Dudai Y (2004) Amygdalar circuits required for either consolidation or extinction of taste aversion memory are not required for reconsolidation. Eur J Neurosci 19:1115–1118

Bailey CH, Bartsch D, Kandel ER (1996) Toward a molecular definition of long-term memory storage. Proc Natl Acad Sci USA 93:13445–13452

Bebien M, Salinas S, Becamel C, Richard V, Linares L, Hipskind RA (2003) Immediate-early gene induction by the stresses anisomycin and arsenite in human osteosarcoma cells involves MAPK cascade signaling to Elk-1, CREB and SRF. Oncogene 22:1836–1847

Biedenkapp JC, Rudy JW (2004) Context memories and reactivation: constraints on the reconsolidation hypothesis. Behav Neurosci 118:956–964

Bourtchouladze R, Abel T, Berman N, Gordon R, Lapidus K, Kandel ER (1998) Different training procedures recruit either one or two critical periods for contextual memory consolidation, each of which requires protein synthesis and PKA. Learn Mem 5:365–374

Bouton ME, Moody EW (2004) Memory processes in classical conditioning. Neurosci Biobehav Rev 28:663–674.

Bozon B, Davis S, Laroche S (2003) A requirement for the immediate early gene zif268 in reconsolidation of recognition memory after retrieval. Neuron 40:695–701

Cahill L, Pham CA, Setlow B (2000) Impaired memory consolidation in rats produced with beta-adrenergic blockade. Neurobiol Learn Mem 74:259–266

Calcagnetti DJ, Schechter MD (1993) Extinction of cocaine-induced place approach in rats: a validation of the "biased" conditioning procedure. Brain Res Bull 30:695–700

Childress AR, Mozley PD, McElgin W, Fitzgerald J, Reivich M, O'Brien CP (1999) Limbic activation during cue-induced cocaine craving. Am J Psych 156:11–18.

Conklin CA, Tiffany ST (2002) Applying extinction research and theory to cue-exposure addiction treatments. Addiction 97:155–167

Davis HP, Squire LR (1984) Protein synthesis and memory: a review. Psychol Bull 96:518–559.

Davis M (2002) Role of NMDA receptors and MAP kinase in the amygdala in extinction of fear: clinical implications for exposure therapy. Eur J Neurosci 16:395–398

de Hoz L, Martin SJ, Morris RG (2004) Forgetting, reminding, and remembering: the retrieval of lost spatial memory. PLoS Biol 2:E225

Debiec J, LeDoux JE (2004) Disruption of reconsolidation but not consolidation of auditory fear conditioning by noradrenergic blockade in the amygdala. Neuroscience 129:267–272

Debiec J, LeDoux JE, Nader K (2002) Cellular and systems reconsolidation in the hippocampus. Neuron 36:527–538

Debiec J, Doyere V, Nader K, LeDoux JE (2006) Directly reactivated, but not indirectly reactivated, memories undergo reconsolidation in the amygdala. Proc Natl Acad Sci USA 103:3428–3433

Di Ciano P, Everitt BJ (2004) Conditioned reinforcing properties of stimuli paired with self-administered cocaine, heroin or sucrose: implications for the persistence of addictive behaviour. Neuropharmacology 47:202–213

Dudai Y (2004) The neurobiology of consolidations, or, how stable is the engram? Ann Rev Psychol 55:51–86

Dudai Y (2006) Reconsolidation: the advantage of being refocused. Curr Opin Neurobiol 16:174–178

Dudai Y, Eisenberg M (2004) Rites of passage of the engram: reconsolidation and the lingering consolidation hypothesis. Neuron 44:93–100

Duvarci S, Nader K (2004) Characterization of fear memory reconsolidation. J Neurosci 24:9269–9275

Eisenberg M, Dudai Y (2004) Reconsolidation of fresh, remote, and extinguished fear memory in medaka: old fears don't die. Eur J Neurosci 20:3397–3403

Eisenberg M, Kobilo T, Berman DE, Dudai Y (2003) Stability of retrieved memory: Inverse correlation with trace dominance. Science 301:1102–1104

Fonseca R, Nagerl UV, Bonhoeffer T (2006) Neuronal activity determines the protein synthesis dependence of long-term potentiation. Nature Neurosci 9:478–480

Frenkel L, Maldonado H, Delorenzi A (2005) Memory strengthening by a real-life episode during reconsolidation: an outcome of water deprivation via brain angiotensin II. Eur J Neurosci 22:1757–1766

Gainutdinova TH, Tagirova RR, Ismailova AI, Muranova LN, Samarova EI, Gainutdinov KL, Balaban PM (2005) Reconsolidation of a context long-term memory in the terrestrial snail requires protein synthesis. Learn Mem 12:620–625

Garavan H, Pankiewicz J, Bloom A, Cho JK, Sperry L, Ross TJ, Salmeron BJ, Risinger R, Kelley D, Stein EA (2000) Cue-induced cocaine craving: neuroanatomical specificity for drug users and drug stimuli. Am J Psych 157:1789–1798

Grant S, London ED, Newlin DB, Villemagne VL, Liu X, Contoreggi C, Phillips RL, Kimes AS, Margolin A (1996) Activation of memory circuits during cue-elicited cocaine craving. Proc Natl Acad Sci USA 93:12040–12045

Guzowski JF (2002) Insights into immediate-early gene function in hippocampal memory consolidation using antisense oligonucleotide and fluorescent imaging approaches. Hippocampus 12:86–104

Guzowski JF, McGaugh JL (1997) Antisense oligodeoxynucleotide-mediated disruption of hippocampal cAMP response element binding protein levels impairs consolidation of memory for water maze training. Proc Natl Acad Sci USA 94:2693–2698

Guzowski JF, Lyford GL, Stevenson GD, Houston FP, McGaugh JL, Worley PF, Barnes CA (2000) Inhibition of activity-dependent arc protein expression in the rat hippocampus impairs the maintenance of long-term potentiation and the consolidation of long-term memory. J Neurosci 20:3993–4001

Hall J, Thomas KL, Everitt BJ (2000) Rapid and selective induction of BDNF expression in the hippocampus during contextual learning. Nature Neurosci 3:533–535

Hall J, Thomas KL, Everitt BJ (2001) Cellular imaging of zif268 expression in the hippocampus and amygdala during contextual and cued fear memory retrieval: Selective activation of hippocampal CA1 neurons during the recall of contextual memories. J Neurosci 21:2186–2193

Hernandez PJ, Kelley AE (2004) Long-term memory for instrumental responses does not undergo protein synthesis-dependent reconsolidation upon retrieval. Learn Mem 11:748–754

Inda MC, Delgado-Garcia JM, Carrion AM (2005) Acquisition, consolidation, reconsolidation, and extinction of eyelid conditioning responses require de novo protein synthesis. J Neurosci 25:2070–2080

Iordanov MS, Pribnow D, Magun JL, Dinh TH, Pearson JA, Chen SL, Magun BE (1997) Ribotoxic stress response: activation of the stress-activated protein kinase JNK1 by inhibitors of the peptidyl transferase reaction and by sequence-specific RNA damage to the alpha-sarcin/ricin loop in the 28S rRNA. Mol Cell Biol 17:3373–3381

Judge ME, Quartermain D (1982) Characteristics of retrograde amnesia following reactivation of memory in mice. Physiol Behav 28:585–590

Kelly A, Laroche S, Davis S (2003) Activation of mitogen-activated protein kinase/extracellular signal-regulated kinase in hippocampal circuitry is required for consolidation and reconsolidation of recognition memory. J Neurosci 23:5354–5360

Kida S, Josselyn SA, de Ortiz SP, Kogan JH, Chevere I, Masushige S, Silva AJ (2002) CREB required for the stability of new and reactivated fear memories. Nature Neurosci 5:348–355.

Lattal KM, Abel T (2004) Behavioral impairments caused by injections of the protein synthesis inhibitor anisomycin after contextual retrieval reverse with time. Proc Natl Acad Sci USA 101:4667–4672

Lechner HA, Squire LR, Byrne JH (1999) 100 years of consolidation–remembering Muller and Pilzecker. Learn Mem 6:77–87

Ledgerwood L, Richardson R, Cranney J (2003) Effects of D-cycloserine on extinction of conditioned freezing. Behav Neurosci 117:341–349

Lee JLC, Everitt BJ, Thomas KL (2004) Independent cellular processes for hippocampal memory consolidation and reconsolidation. Science 304:839–843

Lee JLC, Di Ciano P, Thomas KL, Everitt BJ (2005) Disrupting reconsolidation of drug memories reduces cocaine seeking behavior. Neuron 47:795–801

Lee JLC, Milton AL, Everitt BJ (2006a) Cue-induced cocaine seeking and relapse are reduced by disruption of drug memory consolidation. J. Neurosci.26: 5881–5887

Lee JLC, Milton AL, Everitt BJ (2006b) Reconsolidation and extinction of conditioned fear: inhibition and potentiation. J. Neurosci. 26: 10051–10056

Lewis DJ (1979) Psychobiology of active and inactive memory. Psychol Bull 86:1054–1083

Litvin OO, Anokhin KV (2000) Mechanisms of memory reorganization during retrieval of acquired behavioral experience in chicks: the effects of protein synthesis inhibition in the brain. Neurosci Behav Physiol 30:671–678

Mactutus CF, Riccio DC, Ferek JM (1979) Retrograde amnesia for old (reactivated) memory: some anomalous characteristics. Science 204:1319–1320

Malkani S, Wallace KJ, Donley MP, Rosen JB (2004) An egr-1 (zif268) antisense oligodeoxynucleotide infused into the amygdala disrupts fear conditioning. Learn Mem 11:617–624

McGaugh JL (1966) Time-dependent processes in memory storage. Science 153:1351–1358

McGaugh JL (2000) Neuroscience - memory - a century of consolidation. Science 287:248–251

Merlo E, Freudenthal R, Maldonado H, Romano A (2005) Activation of the transcription factor NF-kappaB by retrieval is required for long-term memory reconsolidation. Learn Mem 12:23–29

Milekic MH, Alberini CM (2002) Temporally graded requirement for protein synthesis following memory reactivation. Neuron 36:521–525

Milekic MH, Brown SD, Castellini C, Alberini CM (2006) Persistent disruption of an established morphine conditioned place preference. J Neurosci 26:3010–3020

Miller CA, Marshall JF (2005) Molecular substrates for retrieval and reconsolidation of cocaine-associated contextual memory. Neuron 47:873–884

Millin PM, Moody EW, Riccio DC (2001) Interpretations of retrograde amnesia: old problems redux. Nature Rev Neurosci 2:68–70

Misanin JR, Miller RR, Lewis DJ (1968) Retrograde amnesia produced by electroconvulsive shock after reactivation of a consolidated memory trace. Science 160:554–555

Nader K (2003) Memory traces unbound. Trends Neurosci 26:65–72

Nader K, Schafe GE, Le Doux JE (2000) Fear memories require protein synthesis in the amygdala for reconsolidation after retrieval. Nature 406:722–726

Nader K, Hardt O, Wang SH (2005) Response to Alberini: right answer, wrong question. Trends Neurosci 28:346–347

Pedreira ME, Maldonado H (2003) Protein synthesis subserves reconsolidation or extinction depending on reminder duration. Neuron 38:863–869

Pedreira ME, Perez-Cuesta LM, Maldonado H (2002) Reactivation and reconsolidation of long-term memory in the crab Chasmagnathus: protein synthesis requirement and mediation by NMDA-type glutamatergic receptors. J Neurosci 22:8305–8311

Pedreira ME, Perez-Cuesta LM, Maldonado H (2004) Mismatch between what is expected and what actually occurs triggers memory reconsolidation or extinction. Learn Mem 11:579–585

Power AE, Berlau DJ, McGaugh JL, Steward O (2006) Anisomycin infused into the hippocampus fails to block "reconsolidation" but impairs extinction: The role of re-exposure duration. Learn Mem 13:27–34

Przybyslawski J, Sara SJ (1997) Reconsolidation of memory after its reactivation. Behav Brain Res 84:241–246

Przybyslawski J, Roullet P, Sara SJ (1999) Attenuation of emotional and nonemotional memories after their reactivation: role of beta adrenergic receptors. J Neurosci 19:6623–6628

Rescorla RA, Wagner AR (1972) A theory of Pavlovian conditioning: variations in the effectiveness of reinforcement and nonreinforcement. In: Black AH and Proteasy WF (eds.) Classical onditioning II: urrent research and theory. New York: Appleton-Century-Crofts, pp 64–99

Ressler KJ, Rothbaum BO, Tannenbaum L, Anderson P, Graap K, Zimand E, Hodges L, Davis M (2004) Cognitive enhancers as adjuncts to psychotherapy: use of D-cycloserine in phobic individuals to facilitate extinction of fear. Arch Gen Psych 61:1136–1144

Routtenberg A, Rekart JL (2005) Post-translational protein modification as the substrate for long-lasting memory. Trends Neurosci 28:12–19

Rudy JW, Biedenkapp JC, Moineau J, Bolding K (2006) Anisomycin and the reconsolidation hypothesis. Learn Mem 13:1–3

Salinska E, Bourne RC, Rose SP (2004) Reminder effects: the molecular cascade following a reminder in young chicks does not recapitulate that following training on a passive avoidance task. Eur J Neurosci 19:3042–3047

Sangha S, Scheibenstock A, Lukowiak K (2003) Reconsolidation of a long-term memory in Lymnaea requires new protein and RNA synthesis and the soma of right pedal dorsal 1. J Neurosci 23:8034–8040

Sangha S, Varshney N, Fras M, Smyth K, Rosenegger D, Parvez K, Sadamoto H, Lukowiak K (2004) Memory, reconsolidation and extinction in Lymnaea require the soma of RPeD1. Adv Exp Med Biol 551:311–318

Sara SJ (2000) Strengthening the shaky trace through retrieval. Nature Rev Neurosci 1:212–213

Schafe GE, LeDoux JE (2000) Memory consolidation of auditory pavlovian fear conditioning requires protein synthesis and protein kinase A in the amygdala. J Neurosci 20:RC96

Schneider AM, Sherman W (1968) Amnesia: a function of the temporal relation of footshock to electroconvulsive shock. Science 159:219–222

Schultz W, Dickinson A (2000) Neuronal coding of prediction errors. Annu Rev Neurosci 23:473–500

Sherry JM, Hale MW, Crowe SF (2005) The effects of the dopamine D1 receptor antagonist SCH23390 on memory reconsolidation following reminder-activated retrieval in day-old chicks. Neurobiol Learn Mem 83:104–112

Stollhoff N, Menzel R, Eisenhardt D (2005) Spontaneous recovery from extinction depends on the reconsolidation of the acquisition memory in an appetitive learning paradigm in the honeybee (Apis mellifera). J Neurosci 25:4485–4492

Strekalova T, Zorner B, Zacher C, Sadovska G, Herdegen T, Gass P (2003) Memory retrieval after contextual fear conditioning induces c- Fos and JunB expression in CA1 hippocampus. Genes Brain Behav 2:3–10

Suzuki A, Josselyn SA, Frankland PW, Masushige S, Silva AJ, Kida S (2004) Memory reconsolidation and extinction have distinct temporal and biochemical signatures. J Neurosci 24:4787–4795

Taubenfeld SM, Milekic MH, Monti B, Alberini CM (2001) The consolidation of new but not reactivated memory requires hippocampal C/EBP beta. Nature Neurosci 4:813–818

Torras-Garcia M, Lelong J, Tronel S, Sara SJ (2005) Reconsolidation after remembering an odor-reward association requires NMDA receptors. Learn Mem 1:18–22

Tronel S, Milekic MH, Alberini CM (2005) Linking new information to a reactivated memory requires consolidation and not reconsolidation mechanisms. PLoS Biol 3:e293

Tronson NC, Wiseman SL, Olausson P, Taylor JR (2006) Bidirectional behavioral plasticity of memory reconsolidation depends on amygdalar protein kinase A. Nature Neurosci 9:167–169

Tzschentke TM, Schmidt WJ (1995) N-methyl-D-aspartic acid-receptor antagonists block morphine-induced conditioned place preference in rats. Neurosci Lett 193:37–40

Valjent E, Corbille AG, Bertran-Gonzalez J, Herve D, Girault JA (2006) Inhibition of ERK pathway or protein synthesis during reexposure to drugs of abuse erases previously learned place preference. Proc Natl Acad Sci USA 103:2932–2937

von Hertzen LS, Giese KP (2005) Memory reconsolidation engages only a subset of immediate-early genes induced during consolidation. J Neurosci 25:1935–1942

Walker DL, Ressler KJ, Lu KT, Davis M (2002) Facilitation of conditioned fear extinction by systemic administration or intra-amygdala infusions of D-cycloserine as assessed with fear-potentiated startle in rats. J Neurosci 22:2343–2351

Walker MP, Brakefield T, Hobson JA, Stickgold R (2003) Dissociable stages of human memory consolidation and reconsolidation. Nature 425:616–620

Wang SH, Ostlund SB, Nader K, Balleine BW (2005) Consolidation and reconsolidation of incentive learning in the amygdala. J Neurosci 25:830–835

The Organizing Principles of Real-Time Memory Encoding: Neural Clique Assemblies and Universal Neural Codes

Joe Z. Tsien[1]

Summary

Recent identification of network-level coding units, termed neural cliques, in the hippocampus has allowed real-time patterns of memory traces to be mathematically described, directly visualized, and dynamically deciphered. Those memory coding units are functionally organized in a categorical and hierarchical manner, suggesting that internal representations of external events in the brain are achieved not by recording exact details of those events but rather by re-creating its own selective pictures based on cognitive importance. These neural clique-based, hierarchical-extraction and parallel-binding processes enable the brain to acquire not only large storage capacity but also abstraction and generalization capabilities. In addition, activation patterns of the neural clique assemblies can be converted to strings of binary codes that would permit universal categorizations of the brain's internal representations across individuals and species.

Neuroscientists try to decipher the brain's neural codes by searching for reliable correlations between firing patterns of neurons and behavioral functions (Adrian 1926; Gross et al. 1972; Fuster 1973; Funahashi et al. 1989; Zhou and Fuster 1996). As early as the 1920s, Edgar Adrian in his pioneering recording showed that the firing rate of a frog muscle's stretch receptor increases as a function of the weights on the muscle (Adrian 1926), suggesting that information is conveyed by specific firing patterns of neurons. Two leading neural coding theories can be found in the literature, namely, a "rate code" and a "temporal code" (Barlow 1972; Van Rullen and Thorpe 2001; Softky 1995; Eggermont 1998). In the rate code, all the information is conveyed in the changes of the firing of the cell. In the temporal code, information is also conveyed in the precise inter-spike intervals. However, due to a large amount of response variability at the single neuron level in the brain, even in response to an identical stimulus (Bialek and Rieke 1992; Fenton and Muller 1998), those two types of single neuron-based decoding schemes often produce significant errors in predictions about the stimulus identities or external information.

Changes in discharge frequency or latencies of neurons upon external stimulations are well known (Gross et al. 1972; Fuster 1973; O'Keefe and Dostrovsky 1971; Miller et al. 1993; Thompson 2005; Leutgeb et al. 2005). To explain how memory might be encoded beyond the level of synapses, Hebb (1949) postulated that information processing

---

[1] Center for Systems Neurobiology, Departments of Pharmacology and Biomedical Engineering, Boston University, Office L-601, 715 Albany Street, Boston, MA 02118, USA
jtsien@bu.edu

Bontempi et al.
Memories: Molecules and Circuits
© Springer-Verlag Berlin Heidelberg 2007

in the brain might involve the coordinated activity of large numbers of neurons, or cell assemblies. This notion, although rather vague, makes good sense both from the computational and cellular perspectives (Wigstrom and Gustafsson 1985; Bliss and Collingridge 1993; Tsien 2000; Abbott and Sejnowski 1999; Shamir and Sompolinsky 2004; Sanger 2003). The major challenge to date has been to identify the actual patterns of activities of a large neuronal population during cognition and then to extract the network-level organizing mechanisms that enable the brain to achieve its real-time encoding, processing, and execution of cognitive information.

Measuring and visualizing network-level memory traces in the hippocampus

We have recently developed a large-scale ensemble recording technique in mice and used novel categorical episodic memory paradigms and powerful mathematical tools to investigate the above questions (Lin et al. 2005). The large-scale ensemble recordings have revealed that various robust episodic experiences, such as free-fall inside a plunging elevator (drop), a sudden gush of air to the back of an animal's back (air blow), or an earthquake-like shake of the home cage (Earthquake), evoked diverse changes in the firing of some of the hippocampal CA1 cells (Fig. 1).

Pattern-classification algorithms, namely, multiple discriminant analysis (MDA) and principle component analysis (PCA), reveal that various startle-triggered ensemble responses of CA1 neurons form distinct patterns in a low-dimensional encoding subspace (Fig. 2). Further application of a sliding-window technique to MDA or PCA methods enables us to directly visualize and dynamically monitor real-time network memory encoding patterns (Fig. 2A).

Interestingly, post-event processing of newly formed ensemble patterns can also be directly detected and precisely quantified (Fig. 2B,C). These spontaneous reactivations of memory traces, represented by dynamic trajectories with similar geometric shapes but smaller amplitudes, seem to occur at intervals ranging from several seconds to minutes after a discrete startling event. The existence of these reactivations suggests that the memory formation is a highly dynamic process and that they may play a crucial role in the immediate post-learning fixation of newly formed memory traces (Fig. 2B,C). Previous studies, based on the comparison of firing covariance value of place cells with overlapping fields between the running sessions and the post-running sleep period, have indicated that place cells participate in reactivations during sleep (Wilson and McNaughton 1994). The detection of awake-state re-activations of memory encoding patterns immediately following the startling events is generally consistent with those interpretations and further illustrates the unprecedented sensitivity of this new decoding method. Thus, the combination of large-scale recording and new decoding algorithms begins to open a door for the direct visualization and quantitative measurement of network-level memory traces and their dynamic temporal evolution.

Identification of neural cliques as real-time memory coding units

To further identify the internal structures underlying the real-time memory encoding, we have employed the agglomerative hierarchical clustering and sequential MDA

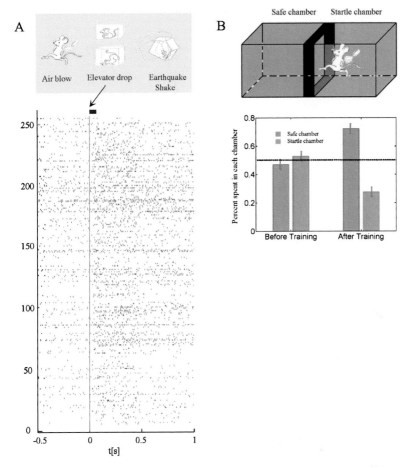

Fig. 1. Large-scale simultaneous monitoring of 260 CA1 cells in mice during various mnemonic startling episodes. (**A**) A set of new behavioral paradigms that involve categorical variables is used to create discrete episodic memories for mice: a sudden blow of air to the animal's back (in mimicking an owl attack from sky); a short vertical freefall inside a plunging small elevator, and an unexpected brief earthquake-like shaking of the mouse's cage. (**B**) The formation of robust memory about the startling events can be assessed by conditioned place conditioning paradigm. The mice spent equal amount of time (~ 50%) between the safe chamber and the startle chamber during the pre-training session prior to startle conditioning. However, after startle conditioning, the mice spent significantly more time in the unconditioned (safe) chamber, as shown in the three-hour retention test (*red bar*, 130.1 ± 5.8 s out of the total 180 seconds, p < 0.0005; numbers of mice: n = 14). The figure is adopted from Lin et al. (2005). A spike raster of 260 simultaneously recorded single units from mouse-A during a period of 0.5 s prior to and 1 s after the occurrence of single startling episodes of Elevator Drop is presented. (t = 0 marked with *vertical red line*). *X-axis*: Time scale (seconds); *Y-axis*: the numbers of simultaneously recorded single units (n = 260). The startle stimulus duration is indicated as a *bar next to the vertical red line above the spike raster*

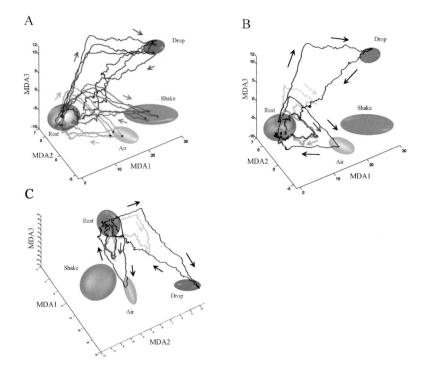

Fig. 2. Classification, visualization, and dynamic decoding of CA1 ensemble representations of startle episodes by Multiple Discriminant Analysis (MDA). (**A**) Ensemble patterns during awake rest (*dots, yellow ellipsoid*), Air-blow (*circles, green ellipsoid*), Drop (*triangles, blue ellipsoid*) and Earthquake (*stars, magenta ellipsoid*) events are shown in a three-dimensional sub-encoding space obtained using MDA for a mouse in which 260 CA1 cells were simultaneously recorded; MDA1-3 denote the discriminant axes. Three representative dynamic trajectories of network patterns during the encoding of each type of startling event are shown. (**B**) Dynamic monitoring of post-learning spontaneous reactivations of network traces during and after the actual startling events. The three-dimensional subspace trajectories of the ensemble encoding patterns during Drop and Air-blow episodes in the same mouse are shown. The initial response to an actual Air-blow or Drop event (*black lines*) is followed by spontaneous reactivations (*red and green lines* for two air-blow reactivations, and *purple line* for Drop pattern reactivation), characterized by co-planar, geometrically similar lower amplitude trajectories (*directionality indicated by arrows*). (**C**) The same trajectories of reactivation traces from a different orientation after a three-dimensional rotation show that the trajectories are highly specific towards their own startle clusters. These post-learning dynamic trajectories are typically smaller in amplitude and take place without any time compression, and the numbers of reactivations within the initial several minutes seem to be in the range of one to five, with random intervals

methods (Lin et al. 2005). These analyses reveal that the network-encoding power is actually derived from a set of functional coding units, termed neural cliques – a group of neurons with a similar response property and selectivity – in the CA1 cell population (Box 1). For example, the "general startle neural clique" consists of individual cells capable of responding to all types of startling stimuli, including the elevator drop,

A

$$f(t) = h(\sum_{n=1}^{N} w_n R_n(t))$$

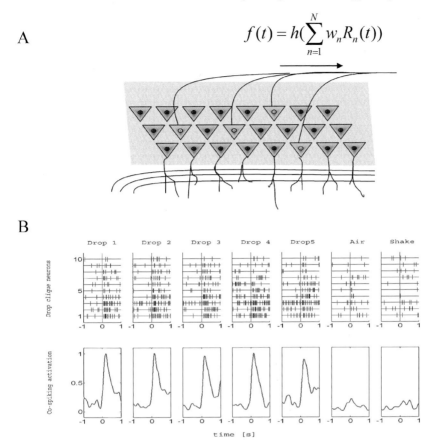

B

Fig. 3. Real-time encoding robustness of memory coding units is achieved through "collective co-spiking" of its individual members within each neural clique. (A) Illustration of a neural clique in the CA1 network. The activation function of the clique to drive the downstream clique can be mathematically described (equation is listed on *top*) as a threshold function of the integrated inputs from the upstream clique members, where R is input, W is the weighting factor, h is the threshold function (e.g., sigmoid), and t is time. (B) Spike rasters and weighted responses of the top 10 neurons within the Drop-specific neural clique (listed in Y-axis) during five elevator-drop events (1 s before and after the startle, X-axis) are shown as an example. Although responses of the individual member (neuron) are quite variable from trial to trial, the consistency and specificity of the collective co-spiking of the clique responses is evident from each Drop episode (five episodes are listed). The Drop-specific neural clique exhibited no significant responses to Air-blow or Shake episodes (last two insets on the *right*). Robust co-spiking of membership neurons in the cliques is also preserved at the finer time-scale (20–30 ms)

earthquake, and air blow, whereas the "sub-general startle cliques" are neural groups that respond to a combination of two types of, but not all, startling events. In addition, there are neuron groups that exhibit high specificity towards one specific type of startling event, such as elevator drop (the drop-specific neural clique), earthquake

Box 1. Categorical and hierarchical organization of the memory-encoding neural clique assembly.

The major organizing principle of network-level memory encoding is that memory coding units are organized in a categorical and hierarchical manner (Lin et al. 2006). For each neural clique assembly underlying the encoding of any given startling episode, there is an invariant internal organization, termed the feature pyramid, with the neural cliques corresponding to the general features at the bottom and neural cliques corresponding to the highly specific features at the top. Such a feature pyramid is evident from the hierarchical clustering analysis of responses of a total of 757 CA1 neurons from four mice to the three different types of startling episode, which reveals the existence of seven major neural cliques (**A**): General startle clique, sub-general startle cliques (Drop-Shake clique, Air blow-Drop cliques, Shake-Air blow clique), startle type-specific cliques (Drop-specific clique, Shake-specific clique, and Air blow-specific clique), and startle context-specific clique (Air- blow in context A-specific clique, Air-blow in context B-specific clique, Drop in Elevator A-specific clique, and Drop in Elevator B-specific clique). Non-responsive units are grouped in the bottom half. The *color scale bar* indicates the normalized response magnitude (1 to 7). A similar cluster is also observed from the simultaneously recorded CA1 populations (Lin et al. 2005).

It is clearly evident that the memory coding units are organized in a hierarchical and categorical fashion (**B**), and any given startling episode is encoded by a combinatorial assembly of a series of neural cliques, invariantly consisting of the general startle clique, sub-general startle clique, startle identity-specific clique, and context-specific startle clique. In this feature pyramid of the encoding clique assembly, the neural clique representing the most general, abstract features (to all categories) is at the bottom and it forms a common building block for all types of startle event encoding. The next layer of the pyramid is made up of neural cliques responding to less general features (covering multiple, but not all, categories); these sub-general cliques are present in a subset of the neural clique assemblies. As one moves up this encoding feature pyramid, neural cliques become more and more specific. The neural clique at the top of the pyramid encodes the most specific and highly discriminate features, thereby defining a particular event or experience. Please note that the number of neurons for each clique does not necessarily correspond to its position in the feature pyramid. In other words, the neural clique encoding the general features does not necessarily have more neurons than the neural cliques encoding more specific features. This invariant feature structure within each neural clique assembly encoding startling episodes reveals four organizing principles for memory encoding in the brain: 1) the memory system employs a categorical and hierarchical architecture in organizing memory coding units; 2) the categorical and hierarchical organization of memory coding units suggests that the internal representations of external events in the memory space is achieved by re-creating its own selective picture, a picture determined largely by what is important for survival and behavioral adaptation; 3) through combinatorial and self-organizing processes, neural cliques can generate vast numbers of unique assembly patterns, thereby providing a network-level mechanism capable of encoding potentially infinite numbers of episodic events; and 4) the neural clique-based hierarchical extraction and parallel binding in the memory system can also enable higher cognition, such as abstraction and generalization capacities, to emerge during the process

(earthquake-specific neural clique), or sudden air-blow events (air puff-specific neural clique).

One can mathematically evaluate the contribution of these neural cliques to the CA1 representations by repeating the MDA analysis while sequentially adding clique members to an initial set of non-responsive neurons. For example, a random selection of 40 non-responsive cells as an initial set provides no discriminating power, yielding only overlapping representations (Lin et al. 2005). In contrast, inclusion of the 10

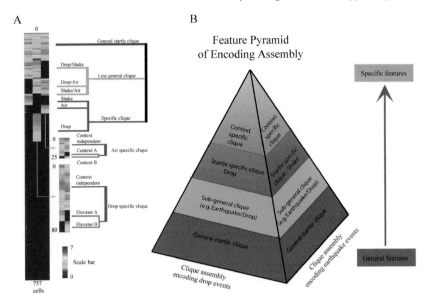

A

B

Feature Pyramid
of Encoding Assembly

most responsive cells from the general startle clique leads to good separation between the rest cluster and the startle clusters, but not among startle clusters. The selective discrimination of "drop" startles is obtained by the addition of as few as 10 top neurons from the drop clique. Similarly, the inclusion of 10 air-blow clique and 10 shake clique top neurons subsequently leads to full discrimination between all startle types. Thus, these neural cliques indeed constitute the basic functional coding units for encoding the identity of different startling episodes.

One crucial feature of neural cliques is that the individual neurons belonging to a given clique exhibit "collective co-spiking" temporal dynamics (Fig. 3). This collective co-spiking dynamics among neural clique members enables the memory coding units to achieve real-time, network-level encoding robustness by overcoming the response variability of individual neurons (Fig. 3). Moreover, based on the temporal dynamics, neurons within each clique can be further sub-grouped into the four major subtypes, namely, 1) transient increase, 2) prolonged increase, 3) transient decrease, and 4) prolonged decrease. The existence of four types of neurons can greatly enhance the real-time encoding robustness as well as provide potential means for modifying clique membership via synaptic plasticity. Finally, neural cliques, as network-level functional coding units, should also be less vulnerable to the death of one or a few neurons and, therefore, exhibit graceful degradation should such conditions arise during the aging process or disease states.

Hierarchical organization within memory-encoding neural clique assemblies

Through examining the overall organization of neural clique assembly involved in startle memory encoding, it is clear that the internal CA1 representations of any given startle episode involves a combinatorial set of neural cliques, invariantly consisting of the general startle clique, sub-general startle clique, startle identity-specific clique, and context-specific startle clique (Lin et al. 2005, 2006). Thus, each clique assembly is organized in a categorical hierarchy manner and invariantly consists of a "feature-encoding pyramid" (Lin et al. 2006; Box 1): starting with the neural clique representing the most general and common features (to all categories) at the bottom layer, followed by neural cliques responding to less general features (covering multiple, but not all, common categories), and then moving gradually up towards more and more specific and discriminating features (responding to a specific category), with the most discriminating feature clique (corresponding to context-specificity) on the top of the feature-encoding pyramid.

According to this hierarchical structure of network-level memory encoding (Box 1), the general startle neural clique represents the neurons engaged in the extraction of the common features among various episodes (e.g., encoding abstract and generalized knowledge that "such events are scary and dangerous" by integrating neural inputs from the amygdala). The sub-general neural cliques are involved in identifying sub-common features across a subset of startling episodes (e.g., perhaps, the earthquake and drop-specific clique for encoding the semantic memory of the fact that "those events involve shaking and motion disturbances" by integrating inputs from the vestibular system), whereas the startle identity-specific cliques encode discriminative information about startle types (defining "what type" of event has happened) and the startle context-specific cliques provide an even more specific feature, such as contextual information about "where" a particular startling event has happened.

This invariant feature-encoding pyramid of neural clique assemblies reveals four basic principles for the organization of memory encoding in the brain (Box 1). First, the neural networks in the memory systems employ a categorical and hierarchical architecture in organizing memory coding units. Second, the internal representations of external events in the brain through such a feature-encoding pyramid is achieved not by recording exact details of the external event but rather by re-creating its own selective pictures based on the importance for survival and adaptation. Third, the feature-encoding pyramid structure provides a network mechanism, through a combinatorial and self-organizing process, for creating seemingly unlimited numbers of unique internal patterns, capable of dealing with potentially infinite numbers of behavioral episodes that an animal or human may encounter during its life. Fourth, in addition to its vast memory storage capacity, these neural clique-based, hierarchical-extraction and parallel-binding processes also enable the brain to achieve abstraction and generalization, cognitive functions essential for dealing with complex, ever-changing situations.

The finding that the memory-encoding neural clique assembly appears to invariantly contain the coding units for processing the abstract and generalized information (Lin et al. 2005, 2006) is interesting. It fits well with the anatomical evidence that virtually all of the sensory input that the hippocampus receives arises from higher-order, multimodal cortical regions and the hippocampus has a high degree of sub-regional divergence and convergence at each loop. This unique anatomical layout supports the notion that whatever processing is achieved by the hippocampus in the service of long-term memory formation should have already engaged with fairly abstract, generalized representations of events, people, facts, and knowledge.

The observed feature-encoding pyramid structure of the neural clique assembly is likely to represent a general mechanism for memory encoding across different animal species. For example, single unit recordings in human hippocampus show that some hippocampal cells fire in response to faces or, more selectively, to a certain type of human facial emotions; others seem to exhibit highly selective firing to one specific person (e.g., the "actress Halle Berry cell", which fires selectively to her photo portraits, Cat-woman character, and even a string of her name (Quiroga et al. 2005). Although those cells were not recorded simultaneously, the findings are nonetheless consistent with the general-to-specific feature pyramid structure. In addition, it is also reported that, whereas some place cells in the rat hippocampus exhibit location-specific firing regardless of whether the animals engage in a random forage or goal-oriented food retrieval (or make a left or right turn in a T-maze), others seem to fire selectively at their place fields only in association with a particular kind of experience (Markus et al. 1995; Wood et al. 2000). Thus, those studies also seem to support the existence of a hierarchical structure involved in space coding. Therefore, the hierarchical organization of the neural clique assembly, revealed through large-scale recordings of startling episodes and mathematical analyses, may represent a general feature for memory encoding in the brains. In addition, it further suggests that episodic memory is intimately linked with and simultaneously converted to semantic memory and generalized knowledge.

This form of hierarchical extraction and parallel binding along CNS pathways into the memory and other high cognition systems is fundamentally different from the strategies used in current computers, camcorders, or intelligent machines. These unique design principles allow the brain to extract the commonalities through one or multiple exposures and to generate more abstract knowledge and generalized experiences. Such generalization and abstract representation of behavioral experiences have enabled humans and other animals to avoid the burden of remembering and storing each mnemonic detail. More importantly, by extracting the essential elements and abstract knowledge, animals can apply past experiences to future encounters that share the same essential features but may vary greatly in physical detail. These higher cognitive functions are obviously crucial for the survival and reproduction of animal species.

Universal activation codes for the brain's real-time neural representations across individuals and species

With the identification of the neural clique as a basic coding unit and the feature-encoding pyramid within the clique assemblies, we can further convert (through matrix inversion) those distinct ensemble representations observed in a low-dimensional encoding-subspace into a string of binary activation codes with 1s and 0s (Fig. 3). This binary assignment – 1 for the active state and 0 for the inactive state of neural cliques – is based on the idea that the activity state of a neural clique can be monitored by downstream neurons using a biologically plausible binary activation function (McCulloch and Pitts 1990; also see Fig. 3). This mathematical conversion of the activation patterns of the neural clique assembly into a binary code of 1s and 0s creates a simple and convenient way for universally comparing and categorizing network-level representations from brain to brain (Lin et al. 2006).

Fig. 3. Conversion of activation patterns of neural clique assemblies into a binary code. (**A**) Conversion of the activation state of a neural clique assembly that encodes one type of startling event into binary digits 1 or 0. (**B**) Mathematical transformation of MDA pattern into a startle type-specific binary encoding system. While the MDA method provides an efficient separation of the startle episodes, each of the discriminant axes at those MDA encoding subspaces (on the left) is no longer corresponding to functional meaning. Thus, we used matrix inversion to translate the ensemble patterns into a startle-specific encoding coordinate system (on the right). This is achieved by assigning new positions for the cluster centers so that they are linearly mapped into a "clique-space", where each axis directly corresponds to a particular clique, thus projecting specific activation patterns to 1 and the absence of activation to 0 (*top panel*). This mathematical operation allow us to map the encoding subspace into one where the startle representations can directly correspond to neural clique activity patterns and, subsequently, to translate the collective activity patterns of neural clique assembly into unique and efficient network-level binary activation codes with a string of binary digits (1's and 0's). (**C**) Conversion of activation patterns of multiple neural clique assemblies into real-time binary codes. Responses of neural cliques are illustrated in different colors. The activation function of a given clique at each network level can be mathematically described (Lin et al. 2005). Rows correspond to the different startling episodes, whereas columns indicate the different neural cliques (General startle, Air-blow, Drop, Shake, Air-blow context-A and Drop context-B). The binary activation patterns corresponding to each event can be mathematically converted to a set of binary codes (on the right column, following the defined sequence of the cliques). The clique activation codes are: 110010 for Air-blow in context A; 110000 for Air-blow in context B, 101000 for Drop in elevator-A; 101001 for Drop in elevator B, and 100100 for Shake. This binary code allows us to accurately predict the behavioral experiences by just sliding through the recorded neural population activity and calculating the hit ratio of matching those binary codes with the occurrences of each startling event. Part C is adopted from Lin et al. (2005)

This type of the universal binary code can provide a potentially unifying framework for the study of high cognition. even across animal species. For example, should a mouse, dog, and human all experience a sudden free-fall in a plunging elevator, the activation patterns of the general startle neural clique, drop-specific clique, air-puff clique, and earthquake clique in their brains would produce the identical real-time binary activation code $(1, 1, 0, 0)$, according to the above permutation and arrangement of the coding unit assembly. Yet, since the mouse, dog, and human may perceive other subtle information differently during the incident, the subsequent digits may differ. For example, the dog may sense a trace amount of smell of burning wires, whereas the human may see erratic flicking of elevator buttons, and the mouse may have a flying candy wrapper hit its face. Therefore, the binary activation codes would permit the universal measurement and categorization of neural representations between those three species, with the initial four digits defining the common experience of free falling and the subsequent digits corresponding to different subtle details.

The proposed binary codes, derived from the activation patterns of the neural clique assembly, offer a concise way to universally categorize the neural representations of cognition in the various brains (Lin et al. 2005, 2006). In the meantime, it is important for us to point out the fundamental differences between the neural clique pattern-based brain codes and the nucleotide-based genetic codes. Specifically, the neural clique-based brain codes have at least four distinct properties: 1) un-inheritable: genetic codes are directly transferred through reproduction, whereas brain codes, by and large,

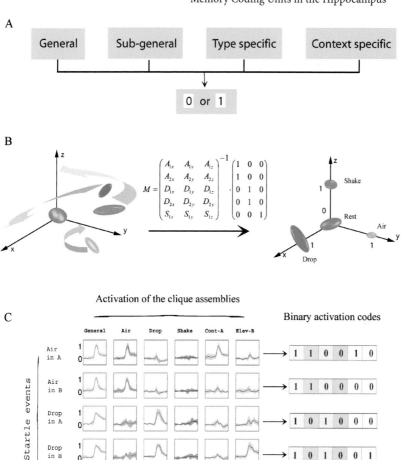

A

| General | Sub-general | Type specific | Context specific |

0 or 1

B

$$M = \begin{pmatrix} A_{1x} & A_{1y} & A_{1z} \\ A_{2x} & A_{2y} & A_{2z} \\ D_{1x} & D_{1y} & D_{1z} \\ D_{2x} & D_{2y} & D_{2y} \\ S_{1x} & S_{1y} & S_{1z} \end{pmatrix}^{-1} \begin{pmatrix} 1 & 0 & 0 \\ 1 & 0 & 0 \\ 0 & 1 & 0 \\ 0 & 1 & 0 \\ 0 & 0 & 1 \end{pmatrix}$$

Shake

Rest

Air

Drop

Activation of the clique assemblies

C   Binary activation codes

| | General | Air | Drop | Shake | Cont-A | Elev-B |

Air in A   1 1 0 0 1 0

Air in B   1 1 0 0 0 0

Drop in A   1 0 1 0 0 0

Drop in B   1 0 1 0 0 1

Shake   1 0 0 1 0 0

Startle events

time [s]

are not inheritable and can only be acquired through experiences (perhaps with the exceptions of those neural codes controlling the primitive functions, such as breathing, heartbeat, and knee-jerk reflex, etc., that may have been genetically programmed); 2) self-organizing: genetic codes act like pre-determined scaffolds, providing blueprints for the development and basic functionality of the organism, whereas brain codes are dynamic and self-organizing, arising out of internal structures and connectivity of neural networks upon behavioral experiences; 3) variable sizes: the numbers of genes are exactly fixed for each individual and species, whereas the number of brain codes is highly variable in each brain; in theory, it is only limited by the network capacity (which is determined by the convergence and divergence in connectivity), as well as the number of behavioral experiences that an individual encounters; and 4) modifiable:

unless mutated, the genetic code remains static, whereas the membership of individual neurons within a given neural clique is modifiable by experience-dependent synaptic plasticity or disease states. Thus, the above features of the brain codes are set apart in a fundamental way from the genetic codes.

The identification of neural cliques as memory coding units in the hippocampus prompts us to entertain the concept that views neural cliques as basic, self-organizing processing units may be applicable to many, if not all, neural networks in the brain. Under this neural clique code model (Box 2), the functionality implemented by neural cliques in a given network depends on the specializations of the corresponding regions. In primary sensory regions, neural cliques in those regions (perhaps, organized in the forms of cortical columns) encode piecemeal information by decomposing external

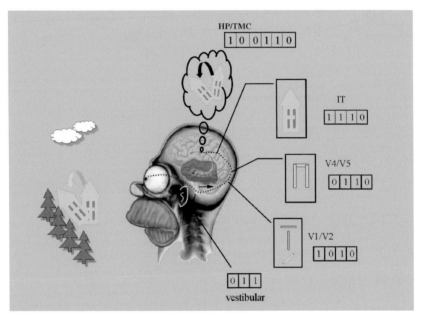

Box 2. Neural clique code-based real-time information processes in the brain.
Through a series of hierarchical-extraction and parallel-binding processes, the brain achieves coherent internal encoding and processing of the external events (Lin et al. 2006). For example, when a person experiences a sudden earthquake, neural cliques in his primary visual cortex encode the decomposed features about edge orientation, movement, and eventually shapes of visual objects, whereas the neural cliques in the vestibular nuclei detect sudden motion disturbances. As information is processed along its pathways into deeper cortex, such as the inferior temporal cortex (IT), neural cliques begin to exhibit complex encoding features such as houses. By the time it reaches high association cortices, such as the hippocampus (HP) and temporal medial cortex (TMC), the neural clique assembly encodes earthquake experience and its location, with a selective set of "what and where" information. At this level, abstract and generalized information such as semantic memories of "the earthquake is dangerous and scary" have emerged. As information is further processed into other cortical regions involving decision-making and motor planning, a series of phased firings among various neural clique assemblies lead to adaptive behaviors such as screaming and running away from the house, or hiding under a dining table

events into various basic features (e.g., the primary visual cortex for detecting edge orientation, color or size of visual objects; the vestibular nuclei for detecting motion, etc. Box 2). As information is further processed along its pathways into deeper regions, neural cliques (although no longer organized in their anatomically distinguishable maps or columns) start to encode more complex features (e.g., shapes and complex objects such as houses and faces in the inferior temporal cortex). By the time they reach high association cortices such as the hippocampus, neural cliques have already contained both specific and generalized mnemonic information about events, places and people with a significant amount of abstraction and generalization (Box 2). Eventually, the brain areas involved in decision-making, executive function and motor planning may start coherent and phased firings among various neural cliques, thereby generating behaviors.

In summary, recent identification of neural cliques as the basic coding units in the brain has provided crucial insights into the network-level organizing principles underlying real-time memory encoding. Those neural cliques are self-organized in a combinatorial fashion to form a memory encoding assembly with an invariant hierarchical structure. This feature-encoding hierarchical structure of the neural clique assembly immediately suggests a network mechanism for the brain to achieve both large memory storage capacity and higher cognitive functions, such as abstraction and generalization.

Acknowledgements. This work was supported by funds from NIMH and NIA, Burroughs Welcome Fund, ECNU Award, and the W.M. Keck Foundation.

References

Abbott LE, Sejnowski TJ (1999) Neural codes and distributed representations. Cambridge, The MIT Press

Adrian EG (1926) The impulses produced by sensory nerve endings: Part 1. J Physiol 61:49–72

Barlow H (1972) Single units and sensation: a doctrine for perceptual psychology? Perception 1:371–394

Bialek W, Rieke F (1992) Reliability and information transmission in spiking neuron. Trends Neurosci 15:428–433

Bliss TV, Collingridge GL (1993) A synaptic model of memory: long-term potentiation in the hippocampus. Nature 361:31–39

Eggermont JJ (1998) Is there a neural code? Neurosci Biobehav Rev 22:355–370

Fenton AA, Muller RU (1998) Place cell discharge is extremely variable during individual passes of the rat through the firing field. Proc Natl Acad Sci USA 95:3182–3187

Funahashi S, Bruce CJ, Goldman-Rakic PS (1989) Mnemonic coding of visual space in the monkey's dorsolateral prefrontal cortex. J Neurophysiol 61:331–349

Fuster JM (1973) Unit activity in prefrontal cortex during delayed-response performance: neuronal correlates of transient memory. J Neurophysiol 36:61–78

Gross CG, Rocha-Miranda CE, Bender DB (1972) Visual properties of neurons in inferotemporal cortex of the macaque. J Neurophysiol 35:96–111

Hebb DO (1949) The organization of behavior. Wiley, New York

Leutgeb S, Leutgeb JK, Moser MB, Moser EI (2005) Place cells, spatial maps and the population code for memory. Curr Opin Neurobiol 15:738–746

Lin L, Osan R, Shoham S, Jin W, Zuo W, Tsien JZ (2005) Identification of network-level coding units for real-time representation of episodic experiences in the hippocampus. Proc Natl Acad Sci USA 102:6125–6130

Lin L, Osan R, Tsien JZ (2006) Organizing principles of real-time memory encoding: Neural clique assemblies and universal neural codes. Trends Neurosci 29:48–56

Markus EJ, Qin YL, Leonard B, Skaggs WE, McNaughton BL, Barnes CA (1995) Interactions between location and task affect the spatial and directional firing of the hippocampal neurons. J Neurosci 15:7079–7094

Miller EK, Li L, Desimone R (1993) Activity of neurons in anterior inferior temporal cortex during a short-term memory task. J Neurosci 13:1460–1478

McCulloch WS, Pitts W (1990) A logical calculus of the ideas immanent in nervous activity. 1943. Bull Math Biol 52:99–115; discussion 173–197

O'Keefe J, Dostrovsky J (1971) The hippocampus as a spatial map. Preliminary evidence from unit activity in the freely-moving rat. Brain Res 34:171–175

Quiroga RQ, Reddy L, Kreiman G, Koch C, Fried I (2005) Invariant visual representation by single neurons in the human brain. Nature 435:1102–1107

Sanger TD (2003) Neural population codes. Curr Opin Neurobiol 13:238–249

Shamir M, Sompolinsky H (2004) Nonliner population codes. Neural Comp 16:1105–1136

Softky WR (1995) Simple codes versus efficient codes. Curr Opin Neurobiol 5:239–247

Thompson RF (2005) In search of memory traces. Annu Rev Psychol 56:1–23

Tsien JZ (2000) Building a brainier mouse. Sci Am 282:62–68

Van Rullen R, Thorpe SJ (2001) Rate-coding versus temporal order coding: what the retinal ganglion cells tell the visual cortex. Neural Compt 13:1255–1283

Wigstrom H, Gustafsson B (1985) On long-lasting potentiation in the hippocampus: a proposed mechanism for its dependence on coincident pre- and postsynaptic activity. Acta Physiol Scand 123:519–522

Wilson MA, McNaughton BL (1994) Reactivation of hippocampal ensemble memories during sleep. Science 265:676–679

Wood E, Dudchenko PA, Robitsek RJ, Eichenbaum H (2000) Hippocampal neurons encode information about different types of memory episodes occurring in the same location. Neuron 27:623–633

Zhou YD, Fuster JM (1996) Mnemonic neuronal activity in somatosensory cortex. Proc Natl Acad Sci USA 93:10533–10537

Making and Retaining New Memories: The Role of the Hippocampus in Associative Learning and Memory

Wendy A. Suzuki

Introduction

The groundbreaking description of the amnesic patient H.M. in the 1950s (Scoville and Milner 1957) demonstrated for the first time that the structures of the medial temporal lobe are critical for our ability to learn and retain new long-term memories for facts and events. This critical form of memory is referred to as declarative memory in humans (Squire et al. 2004) and relational memory in animals (Eichenbaum et al. 1999). The development of powerful animal models of human amnesia in monkeys (Zola-Morgan and Squire 1990; Mishkin 1978; Zola and Squire 2000; Suzuki et al. 1993) and in rodents (Bunsey and Eichenbaum 1993,1995, 1996; Fortin et al. 2002), together with detailed neuroanatomical studies (Suzuki and Amaral 1994a,b; Burwell and Amaral 1998a,b), demonstrated definitively that the key medial temporal lobe structures important for declarative/relational memory include the hippocampus together with the surrounding entorhinal, perirhinal and parahippocampal cortices. While this convergence of studies in humans and animals has provided detailed information about the pattern of memory impairment following a wide range of lesions to the medial temporal lobe, less information is known about how individual cells in the intact medial temporal lobe participate in the acquisition, consolidation or retrieval of various forms of declarative/relational memory.

To address this question, we have used single unit electrophysiological recording techniques to examine the patterns of neural activity in monkeys performing controlled, memory-demanding tasks. We have focused on associative memory, defined as the ability to learn the relationship between unrelated items such as the name of someone you have just met or the shelf in the kitchen where the teapot is stored. Importantly, this form of declarative/relational memory is known to be highly dependent on the integrity of the medial temporal lobe (Brasted et al. 2002, 2003; Rupniak and Gaffan 1987; Murray and Wise 1996; Murray et al. 2000; Wise and Murray 1999).

Figure 1 illustrates the three major stages in the "life" of an associative memory. The first stage of associative memory is acquisition, when the association is encountered for the first time and the relationship between the unrelated elements is first learned. During the consolidation stage, the newly formed associative memory can be strengthened through repetition. However, lesion studies have demonstrated that the memories remain dependent on the medial temporal lobe during this stage (Bayley et al. 2005). If the new associative information is successfully consolidated and it enters

---

[1] Center for Neural Science, New York University, Building Meyer, Room 809, 4 Washington Place, New York, NY 10003, USA
wendy@cns.nyu.edu

Bontempi et al.
Memories: Molecules and Circuits
© Springer-Verlag Berlin Heidelberg 2007

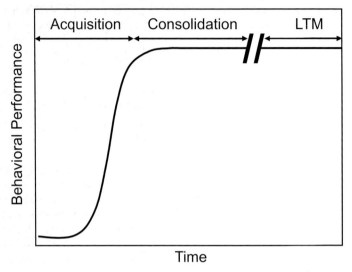

Stages of Memory

Fig. 1. Stages of memory. Illustration of the three major stages of an associative memory, starting with acquisition, when the association is just being learned, consolidation, during which time the memory is susceptible to disruption, and finally the long term memory (LTM) stage, when the memory becomes resistant to damage to the medial temporal lobe

long-term memory, it is no longer susceptible to disruption by damage to the medial temporal lobe, although this structure still may participate in aspects of representation or retrieval (Schacter and Wagner 1999; Schacter et al. 1995, 1996; Nyberg et al. 1996a,b).

We have been particularly interested in the dynamic aspects of associative memory formation, and our first studies focused on the patterns of activity of hippocampal neurons during the early acquisition phase, when new associative memories are first being formed. These studies revealed strong associative learning signals in the monkey hippocampus. We next asked how these associative learning-related signals might be transformed as the initially novel associations became highly familiar with weeks and months of practice. To address this question, we examined the patterns of hippocampal responses to very well-learned associations and compared them to the responses to novel associations. Our data suggest that the hippocampus not only plays a critical role in signaling new associations but also that there is a strong representation of well-learned information in the hippocampus.

The Neural Correlates of Associative Memory Formation

Location-Scene Association Task

To examine the patterns of neural activity during the formation of new associative memories, we recorded the activity of individual hippocampal neurons as monkeys performed a location-scene association task (Wirth et al. 2003). In this task, animals

Location-scene association task

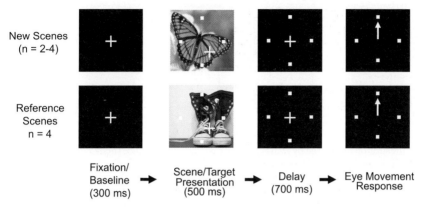

Fig. 2. Location-scene association task. In this task, following fixation, animals are shown a set of four identical visual targets superimposed on a complex visual image (images used in task were all in color). Following a delay interval, during which time the targets remained on the screen but the scene disappeared, the animal was cued to make an eye movement response (illustrated schematically by the white arrow) to one of the targets. Only one of the targets was rewarded for each particular scene. Animals learned by trial and error to associate each new scene with a particular eye movement response. Animals were also shown highly familiar "reference" scenes that they had seen many times before and on which they performed at ceiling levels

learned which one of four identical targets superimposed on a complex visual scene was associated with reward (Fig. 2). The task started with the monkeys fixating on a central fixation point. They were then presented with four identical targets superimposed on a complex visual scene. Following a delay interval where the visual scene disappeared but the targets remained on the screen, the fixation spot was extinguished, which was the animal's cue to make an eye movement to one of the four possible targets. Only one of the targets was rewarded for each new scene. Each day, animals were presented with a random mix of two to four new scenes (each associated with a unique rewarded target location) together with two to four highly familiar "reference" scenes (also associated with four different rewarded target locations). The new location-scene associations were learned in an average of 12 ± 1 trial and animals performed the reference scenes at or near ceiling levels.

Learning-related neural activity in the monkey hippocampus

We hypothesized that cells signaling learning would change their activity in a manner that was closely correlated with the animal's behavioral learning curve for the new scenes, but not for references scenes. Consistent with this hypothesis, we found that 18% of the total population of sampled hippocampal cells (or 28% of the cells that responded selectively during at least one phase of the task) signaled new learning with dramatic changes in firing rates (Fig. 3). This changing neural activity was significantly correlated with the animal's behavioral learning curve. We call these cells "changing" cells. Approximately half the changing cells increased their activity during either the

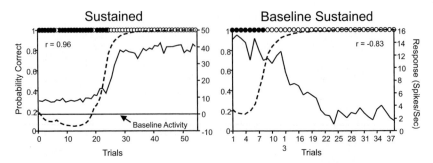

Fig. 3. Illustration of the trial by trial probability correct performance (*dotted line read from the left axis*) as a function of the trial by trial activity of cells during either the scene or delay period of the task (*solid line read from right axis*) for a sustained (*left*) and baseline sustained (*right*) cell. Note the strong positive or negative correlation between neural activity and learning

scene or delay periods of the task correlated with learning, and this change in neural activity was sustained for as long as we were able to hold the cell (sustained changing cells; Fig. 3, right). Some of the sustained changing cells started out with very weak or no response during the early trials of the session and only developed a strong response as the animal learned a particular association.

The remaining half of the changing cells responded robustly and selectivity to a particular scene early in the session before any association was learned. These cells signaled learning by decreasing their firing rate to baseline levels, and this decreasing activity was anti-correlated with learning (baseline sustained changing cells). The responses of both sustained and baseline sustained changing cells were highly selective, in that the changes in neural activity only occurred for a particular learned scene. To determine if these changes in neural activity represent a motor signal, we compared the response of the changing cell to the response of the same cell to the reference scenes with the same rewarded target location (i.e., the same motor response for both new and reference scenes). In no case did the changing cells respond similarly to the references scenes, suggesting the changing signal was not simply signaling the direction of movement. To determine if the changing cells signal learning specific for particular rewarded target locations, we recorded the activity of a changing cell during learning of two different new scenes with the same rewarded target location. Typically, the cell would change in parallel to learning for the first new scene. However, similar changes in activity were never seen for the second new scene with the same rewarded target location. Thus, hippocampal changing cells do not appear to signal new learning in a motor-based or direction-based frame of reference. Instead, these findings suggest that hippocampal cells signal fast associative learning between sensory stimuli and motor responses or target locations. This interpretation is consistent with theories suggesting that the hippocampus plays a fundamental role in forming the random associations or relationships between unrelated items (Eichenbaum et al. 1999). These kinds of simple associations may be critical to build up the more complex associations between the "what", "where" and "when" information that forms the basis of episodic memories.

What does the change in neural activity represent?

Previous studies have shown that neurons in both the perirhinal cortex and area TE signal long-term associations between visual stimuli (visual-visual paired associate memory) by responding similarly to the two items that had been paired in memory (Naya et al. 2003; Sakai and Miyashita 1991). These findings suggested that the learning of the paired associates may have "tuned" or "shaped" the sensory responses of these cells towards a similar response to the two stimuli paired in memory. Consistent with this idea, several other groups demonstrated that perirhinal neurons show a shift in response selectivity during the associative learning process (Erickson and Desimone 1999; Erickson et al. 2000; Messinger et al. 2001). These findings suggested that the striking changes in neural activity observed during the location-scene association task may represent a change in the cell's stimulus selective response properties with learning. To address this possibility, we examined the average response of a single changing cell to all new scenes and reference scenes over the course of learning. The example cell, illustrated in Fig. 4A, did not differentiate between any of the new or reference scenes during the scene period of the task early in the learning session. However, this cell appeared to develop a highly selective response to new scene 2 (black line), which occurred in parallel with learning (thick gray line). To quantify this observation across the changing cells, we measured selectivity using a selectivity index (Moody et al. 1998) to the responses to all new and reference scenes before versus after learning. We analyzed the sustained and baseline sustained changing cells separately. We found that, whereas the sustained cells exhibited a significant increase in selectivity (Fig. 4B) with learning, the baseline sustained cells exhibited a significant decrease in selectivity (Fig. 4C). These findings suggest that hippocampal cells signal new associations with a significant change in their stimulus-selective response properties.

Timing of the changing cells relative to learning

We next focused on understanding the timing of these neural changes relative to behavioral learning. If these changing cells changed slightly before behavioral learning was expressed, this would suggest a role in driving the learning process. Changes in neural activity that occurred after behavioral learning was expressed would be consistent with a role in strengthening of the newly formed association. To address this question, for each cell that changed for a particular condition, we estimated the trial number of learning as well as the trial number of neural change. Figure 5 shows a scatter plot of this comparison for all changing cells in the hippocampus. This plot makes several important points. First, whereas about half of the changing cells changed before or at the same time as learning, the remaining half changed after learning. These early changing cells suggest that the hippocampus may be among the earliest brain structures to signal or drive new associative learning. Second, these data also suggest that, while early learned associations (i.e., associations learned in less than about 15 trials) tended to change after learning, neurons only started to change before behavioral learning if the animal took longer to learn the association. This pattern suggests the possibility that different behavioral (and neural) strategies may be employed early in the session, when the scenes are novel and there is more overall uncertainty associated

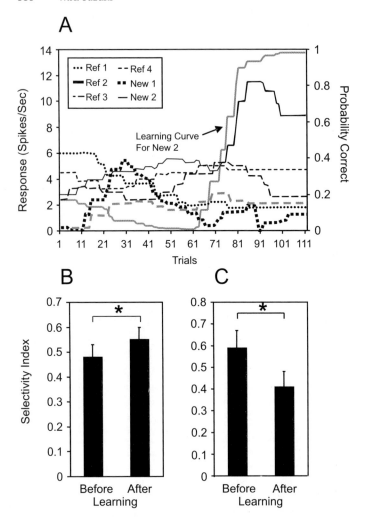

Fig. 4. (**A**) Average response to four reference scenes and two new scenes over the course of the recording session for a sustained changing cell. The learning curve for New Scene 2 is illustrated by the *thick gray line*. (**B**) Graph showing the significant increase in selectivity index for sustained changing cells after learning compared to before learning. (**C**) In contrast, baseline sustained changing cells decreased their selectivity after learning compared to before learning

with them, and later in the session, when the scenes are more familiar and at least one or two have been learned.

To summarize thus far, we have shown that many hippocampal neurons signal learning by changing their stimulus-selective response properties. While some neurons increase their activity with learning, others decrease their activity (and stimulus-selective responses), suggesting that these findings may represent an overall tuning of the network of hippocampal responses with learning. Because these changes occur

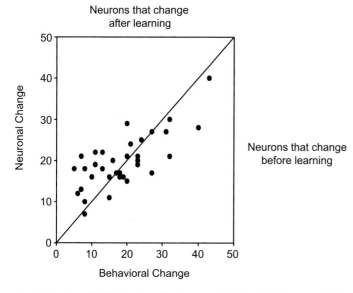

Fig. 5. Scatter plot illustrating the temporal relationship between trial number of behavioral change (i.e., learning) and trial number of neuronal change. Note that about half the cells change before or at the same time as learning whereas the remaining half of the cells change before learning. Moreover, there is a tendency for cells to change after learning if the scene was learned quickly (i.e., faster than about 15 trials) and to change after learning if the scene was learned more slowly

before, at the same time as, and after learning, there may be a gradual recruitment of a network of hippocampal neurons throughout the early formation of new associative memories.

The Neural Correlates of Well-Learned Associations

While the previous section illustrated the dramatic changes that occur in hippocampal activity early in learning, we next asked if this signal might persist in some form as the memory is consolidated to represent very well-learned information. Previous findings have shown that cortical areas exhibited two general categories of long-lasting, experience-dependent plasticity. First, some cells exhibit a gradual decrease in neural activity as an initially novel stimulus becomes familiar (Fahy et al. 1993; Riches et al. 1991; Li et al. 1993). Other studies have reported cortical neurons that increase their stimulus-selective responses to particular, well-learned stimuli (Baker et al. 2002; Kobatake et al. 1998; Logothetis and Pauls 1995; Sigala et al. 2002). While the hippocampus has most commonly been associated with new learning and consolidation of newly learned information, other studies have suggested a role in retrieval of well-learned information (Schacter and Wagner 1999; Schacter et al. 1995, 1996; Nyberg et al., 1996a,b). Indeed the role of the hippocampus in signaling information about well-learned information has not been examined in great detail. To address this question, we examined the responses of hippocampal neurons to the well-learned reference scenes of our location-scene association task (Yanike et al. 2004; see also Fig. 2). In most cases, the animals had between 6 and 22 months of previous experience with the reference scenes before the cells were recorded. Thus, these reference scenes were both well-learned and extensively practiced.

Selectivity for reference vs. novel scenes

Similar to reports for other cortical areas, we found that the selectivity of the hippocampus was modulated by the extensive training with the reference scene compared to the new scenes. Using a selectivity index (SI Moody et al. 1998), we analyzed the responses of hippocampal cells with selective responses during the scene and delay periods of the task, comparing the selectivity to new and reference scenes. We found that the average SI for the well-learned references scenes was significantly higher than the average SI for the new scenes during both the scene period of the task and the delay period of the task. Thus, hippocampal cells respond significantly more selectively to the well-learned references scenes compared to the novel scenes. An example of an individual hippocampal cell with a higher SI to reference scenes compared to new scenes is shown in Fig. 6. This cell responded maximally to one reference scene but did not discriminate between the other new or reference scenes in that day's set.

Time course of the selective response

To further characterize the differences in selective responses to new and reference scenes, we analyzed the temporal dynamics of when the differences in selectivity first appeared. For each cell, we computed the average activity over successive 50-ms bins aligned at the scene onset, and we calculated a selectivity index (depth of tuning index, or DTI; Moody et al. 1998) for each of the successive 50-ms bins throughout the scene and delay periods of the task separately for reference and new scenes. We found that the enhanced selectivity was present as soon as the cells responded to the visual stimuli. Moreover, the enhanced selectivity was maintained throughout the scene and delay periods of the task.

While many previous studies have shown that neocortical cells in sensory areas can shift their sensory tuning with practice, our findings show that hippocampal neurons exhibit this same pattern of increasing selectivity to highly familiar stimuli. These results are particularly surprising given the previous reports that damage to the hippocampus does not impair performance on well-learned associations (Murray and Wise 1996; Wise and Murray 1999). These physiology studies suggest that, despite the fact that other areas may be able to sustain memory for well-learned information, the hippocampus nonetheless participates in the representation of well-learned information. These findings are consistent with a role in retrieval of well-learned information suggested by functional imaging studies in humans (Schacter and Wagner 1999; Schacter et al. 1995, 1996; Nyberg et al. 1996a,b).

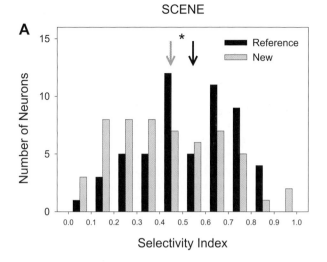

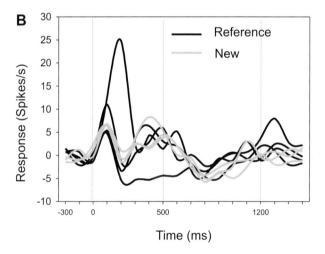

Fig. 6. (**A**) Distribution of selectivity index for all selectively responding hippocampal cells shown separately for new and reference scenes. The average SIs for the reference scene and new scenes were 0.55 and 0.47, respectively. These values are significantly different from each other. (**B**) Response of a single selective hippocampal cell to four reference scenes and three new scenes. This cell responded highly selectively to a particular reference scene. The SI to the references scenes was 0.7 whereas the corresponding value for the new scenes was 0.3

Conclusions

The findings summarized above demonstrate that hippocampal activity is involved in two key stages in the "life" of an associative memory. First, the hippocampus provides a prominent and highly selective signal for the encoding of new associations. A detailed analysis of the temporal relationship of these changing neural signals relative to behavior suggests that hippocampal neurons can signal learning before, at the same time as, and after learning. This finding suggests a role both in driving learning behavior and in a strengthening process. We also showed that these plastic processes are reflected in a shift in the stimulus-selective response properties of the hippocampal cells. The same

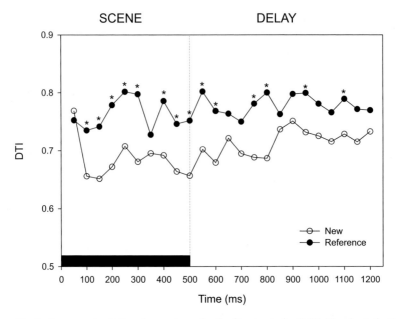

Fig. 7. Average selectivity values using a depth of tuning index (DTI; Moody et al., 1998) plotted for successive 50-ms bin intervals through the scene and delay periods of the task. Asterisks indicate significant differences. *Dark line* on the X-axis shows the time during which the stimulus was shown

population of sampled hippocampal cells also signals very well-learned and extensively practiced associations with a significantly more selective response compared to novel scenes. These results suggest the possibility that the plasticity underlying the changing activity during learning may also underlie the increased selectivity seen in response to well-learned reference scenes. Thus, with extensive daily experience, the striking learning-related plasticity observed during initial learning may eventually evolve into the enhanced selectivity to highly familiar reference scenes. This latter form of selective activity may play a role in the representation of well-learned information as well as in the retrieval process. Future studies will be needed to better define the time course of these effects and the role of the hippocampus in the strengthening or consolidation process that occurs between new learning and long-term memory representation.

References

Baker CL, Behrmann M, Olson CR (2002) Impact of learning on representation of parts and wholes in monkey inferotemporal cortex. Nature Neurosci 5:1210–1216

Bayley PJ, Gold JJ, Hopkins RO, Squire LR (2005) The neuroanatomy of remote memory. Neuron 46:799–810

Brasted PJ, Bussey TJ, Murray EA, Wise SP (2002) Fornix transection impairs conditional visuomotor learning in tasks involving nonspatially differentiated responses. J Neurophysiol 87:631–633

Brasted PJ, Bussey TJ, Murray EA, Wise SP (2003) Role of the hippocampal system in associative learning beyond the spatial domain. Brain 126:1202–1223

Bunsey M, Eichenbaum H (1993) Critical role of the parahippocampal region for paired-associate learning in rats. Behav Neurosci 107:740–747

Bunsey M, Eichenbaum H (1995) Selective damage to the hippocampal region blocks long term retention of a natural and nonspatial stimulus-stimulus association. Hippocampus 5:546–556

Bunsey M, Eichenbaum H (1996) Conservation of hippocampal memory functions in rats and humans. Nature 379:255–257

Burwell RD, Amaral DG (1998a) Cortical afferents of the perirhinal, postrhinal and entorhinal cortices. J Comp Neurol 398:179–205

Burwell RD, Amaral DG (1998b) Perirhinal and postrhinal cortices of the rat: interconnectivity and connections with the entorhinal cortex. J Comp Neurol 391:293–321

Eichenbaum H, Dudchenko P, Wood E, Shapiro M, Tanila H (1999) The hippocampus, memory, and place cells: Is it spatial memory or a memory space? Neuron 23:209–226

Erickson CA, Desimone R (1999) Responses of macaque perirhinal neurons during and after visual stimulus association learning. J Neurosci 19:10404–10416

Erickson CA, Jagadeesh B, Desimone R (2000) Clustering of perirhinal neurons with similar properties following visual experience in adult monkeys. Nature Neurosci 3:1143–1148

Fahy FL, Riches IP, Brown MW (1993) Neuronal activity related to visual recognition memory: long-term memory and the encoding of recency and familiarity information in the primate anterior and medial inferior temporal and rhinal cortex. Exp Brain Res 96:457–472

Fortin NJ, Agster KL, Eichenbaum HB (2002) Critical role of the hippocampus in memory for sequences of events. Nature Neurosci 5:458–462

Kobatake E, Wang G, Tanaka K (1998) Effects of shape-discrimination training on the selectivity of inferotemporal cells in adult monkeys. J Neurophysiol 80:324–330

Li L, Miller EK, Desimone R (1993) The representation of stimulus familiarity in anterior inferior temporal cortex. J Neurophysiol 69:1918–1929

Logothetis NK, Pauls J (1995) Psychophysical and physiological evidence for viewer-centered object representations in the primate. Cereb Cortex 3:270–288

Messinger A, Squire LR, Zola SM, Albright TD (2001) Neuronal representations of stimulus associations develop in the temporal lobe during learning. Proc Natl Acad Sci USA 98:12239–12244

Mishkin M (1978) Memory in monkeys severely impaired by combined but not by separate removal of amygdala and hippocampus. Nature 273:297–298

Moody SL, Wise SP, di Pellegrino G, Zipser DA (1998) A model that accounts for activity in primate frontal cortex during a delayed matching to sample task. J Neurosci 18:399–410

Murray EA, Wise SP (1996) Role of the hippocampus plus subjacent cortex but not amygdala in visuomotor conditional learning in rhesus monkeys. Behav Neurosci 110:1261–1270

Murray EA, Bussey TJ, Wise SP (2000) Role of prefrontal cortex in a network for arbitrary visuomotor mapping. Exp Brain Res 133:114–129

Naya Y, Yoshida M, Miyashita Y (2003) Forward processing of long-term associative memory in monkey inferotemporal cortex. J Neurosci 23:2861–2871

Nyberg L, McIntosh AR, Cabeza R, Habib R, Houle S, Tulving E (1996a) General and specific brain regions involved in encoding and retrieval of events: What, where, and when. Proc Natl Acad Sci USA 93:11280–11285

Nyberg L, McIntosh AR, Houle S, Nilsson LG, Tulving E (1996b) Activation of medial temporal structures during episodic memory retrieval. Nature 380:715–717

Riches IP, Wilson FA, Brown MW (1991) The effects of visual stimulation and memory on neurons of the hippocampal formation and the neighboring parahippocampal gyrus and inferior temporal cortex of the primate. J Neurosci 11:1763–1779

Rupniak NM, Gaffan D (1987) Monkey hippocampus and learning about spatially directed movements. J Neurosci 7:2331–2337

Sakai K, Miyashita Y (1991) Neural organization for the long-term memory of paired associates. Nature 354:152–155

Schacter DL, Wagner AD (1999) Medial temporal lobe activations in fMRI and PET studies of episodic encoding and retrieval. Hippocampus 9:7–24

Schacter DL, Reiman E, Uecker A, Polster MR, Yun LS, Cooper LA (1995) Brain regions associated with retrieval of structurally cohrent visual infomation. Nature 376:587–590

Schacter DL, Alpert NM, Savage CR, Rauch SL, Albert MS (1996) Conscious recollection and the human hippocampal formation: Evidence from positron emission tomography. Proc Natl Acad Sci USA 93:321–325

Scoville WB, Milner B (1957) Loss of recent memory after bilateral hippocampal lesions. J Neurol NeurosurgPsych 20:11–21

Sigala N, Gabbiani F, Logothetis NK (2002) Visual categorization and object representation in monkeys and humans. J Cogn Neurosci 14:187–198

Squire LR, Clark RE, Bailey PJ (2004) Medial temporal lobe function and memory. In: Gazzaniga M (ed) The cognitive neurosciences III. The MIT Press, Cambridge, pp 691–708

Suzuki WA, Amaral DG (1994a) Perirhinal and parahippocampal cortices of the macaque monkey: Cortical afferents. J Comp Neurol 350:497–533

Suzuki WA, Amaral DG (1994b) Topographic organization of the reciprocal connections between monkey entorhinal cortex and the perirhinal and parahippocampal cortices. J Neurosci 14:1856–1877

Suzuki WA, Zola-Morgan S, Squire LR, Amaral DG (1993) Lesions of the perirhinal and parahippocampal cortices in the monkey produce long-lasting memory impairment in the visual and tactual modalities. J Neurosci 13:2430–2451

Wirth S, Yanike M, Frank LM, Smith AC, Brown EN, Suzuki WA (2003) Single neurons in the monkey hippocampus and learning of new associations. Science 300:1578–1581

Wise SP, Murray EA (1999) Role of the hippocampal system in conditional motor learning: Mapping antecedents to action. Hippocampus 9:101–117

Yanike M, Wirth S, Suzuki WA (2004) Representation of well-learned information in the monkey hippocampus. Neuron 42:477–487

Zola SM, Squire LR (2000) The medial temporal lobe and the hippocampus. In: Tulving E, Craik FIM (eds) The Oxford handbook of memory. Oxford: Oxford University Press 485–500

Zola-Morgan S, Squire LR (1990) The neuropsychology of memory. Parallel findings in humans and nonhuman primates. Ann NY Acad Sci 608:434–450

Entorhinal Grid Cells and the Representation of Space

Francesca Sargolini[1] and *Edvard I. Moser*[2]

Summary

The ability to find one's way depends on the brain's ability to integrate information about location, direction and distance. The algorithms responsible for this integration are implemented in a large brain network that includes both hippocampal and parahippocampal cortices, as indicated by the existence of place cells in the hippocampus and head-direction cells in the dorsal presubiculum and a number of other regions. Recent results have pointed to the medial entorhinal cortex (MEC) as a possible site for the dynamic representation of position in animals that move through a two-dimensional environment. Layer II of the MEC contains position-sensitive neurons – grid cells – whose firing fields form a periodic triangular pattern that tiles the entire environment covered by the animal during exploration of an open surface. Grid cells are observed in all principal layers of MEC, but intermingle with head direction cells in layers III, V and VI. The two cell types form a continuous population in which a subset of the neurons, predominantly in layers III and V, have conjunctive grid and head-direction properties. The majority of the cells are modulated by velocity. These observations suggest that, despite the differential hippocampal and neocortical connections of different layers of the MEC, the area operates as an integrated unit, with significant interlaminar interaction between cells with different functional properties. As the animal moves across a surface, activity may be translated across the superficial sheet of grid cells by convergence of position, direction and velocity information in a neural subpopulation with conjunctive firing properties.

Introduction

More than three decades of research have pointed to the hippocampus as a key region of the brain's neural map of the spatial environment (O'Keefe and Nadel 1978). An important milestone was reached when O'Keefe and Dostrovsky (1971) discovered that cells in the hippocampus have spatial receptive fields (place fields), discharging only when the animal was at particular locations in the environment. A number of studies showed that spatial memory was impaired after damage to the hippocampus (Olton et al. 1978; Morris et al. 1982; Nadel 1991), and the hippocampus was found to be activated in normal human subjects as they searched for hidden goals in a spatial

[1] Centre for the Biology of Memory, Norwegian University of Science and Technology, NO-7489 Trondheim, Norway, edvard.moser@cbm.ntnu.no
[2] Laboratoire de Neurobiologie de la Cognition Université deProvence CNRS-UMR6155 Pôle 3C, Case c, 3 Place Victor Hugo 13331 Marseille cedex 03, France

Bontempi et al.
Memories: Molecules and Circuits
© Springer-Verlag Berlin Heidelberg 2007

environment (Maguire et al. 1997, 1998). The animal's position in the external world was shown to be represented in the collective activity of hippocampal neurons; from the simultaneous firing of a cluster of co-localized neurons, it became possible to predict with considerable accuracy the animal's location as it moved through an open field (Wilson and McNaughton 1993). As the years passed, however, accumulating evidence suggested that the functions of place cells extended beyond a specific function in mapping of the physical space, confirming the impression of clinical neuroscientists that hippocampal lesions led to impairment in a much wider spectrum of associative memory functions (Scoville and Milner 1957; Squire et al. 2004). Place cells were shown to respond to a variety of non-spatial sensory inputs (Wood et al. 1999, 2000; Frank et al. 2000; Hollup et al. 2001) and to alternate between multiple representations in the same location, reflecting both salient physical properties of the place and events associated with the place (Bostock et al. 1991; Markus et al. 1995; Kentros et al. 1998; Lever et al. 2002; Leutgeb et al. 2005a,b). Based on these and a number of other observations, most researchers have converged towards the view that the hippocampus has a broad role in the encoding of context-specific or episodic memories, in which spatial location is a critical but non-exclusive part of what is stored in this region (Leutgeb et al. 2005c).

One of the strongest pieces of evidence for a spatial map in the hippocampus was the apparent contrast between the low spatial information of entorhinal neurons and the high spatial information of place cells downstream in the hippocampus (Barnes et al. 1990; Quirk et al. 1992; Frank et al. 2000), which suggested that the location-specific signal was computed internally in the hippocampal circuitry, between the input neurons of the superficial layers of the entorhinal cortex and the place cells of the hippocampus proper. Several studies during the last few years have refuted this view and instead suggested that the animal's current position is computed upstream of the hippocampus in the MEC (Fig. 1), at the intersection between directional input from the presubiculum and visual and movement-related inputs from the postrhinal and retrosplenial cortices (Witter and Amaral 2004). With the aid of small microelectrodes, we have recorded neural activity simultaneously from a number of individually separable neurons in MEC in rats that explore spatial environments of different sizes and shapes. Many principal neurons in layers II and III of MEC were found to have multiple, sharply delineated firing fields that collectively signalled the rat's current position as accurately as place cells in the hippocampus (Fyhn et al. 2004). The sharpest firing fields were observed dorsocaudally in the MEC, in the dorsolateral band, which contains the majority of the projection neurons to the well-defined place cells of the dorsal hippocampus (Witter et al. 1989; Dolorfo and Amaral 1998). Firing fields became gradually wider and more dispersed along a dorsolateral-to-ventromedial axis through the MEC. In the intermediate-to-ventromedial bands, where all of the previous recordings had been made (Quirk et al. 1992; Frank et al. 2000), firing fields were diffuse and covered almost the entire recording environment (Fyhn et al. 2004). No spatial structure could be observed at the ventral end of MEC. Lesion studies confirmed the importance of the dorsolateral band for navigation in spatial laboratory tasks (Steffenach et al. 2005). Localized firing in MEC neurons was maintained without a functional hippocampus, and similar spatially modulated signals were not observed in the postrhinal cortex, from which most of the visuospatial afferent input originates (Witter et al. 1989; Burwell 2000; Witter and Amaral 2004). These findings all suggest that information about current position is, to a large extent, computed intrinsically in the entorhinal cortex.

A

B

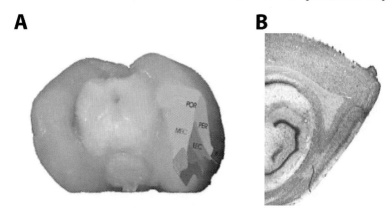

Fig. 1. Entorhinal cortex of the rat. (**A**) Ventral posterior view of the whole rat brain. POR, postrhinal cortex; PER, perirhinal cortex; MEC, medial entorhinal cortex; LEC, lateral entorhinal cortex. (**B**) Sagittal Nissl-stained section of the rat brain. *Red dot* indicates a recording position at the dorsocaudal end of MEC. *Red lines* indicate the dorsal and ventral borders of the MEC. (**A**) is reproduced from Sargolini et al. (2006)

The observation that entorhinal neurons have spatial receptive fields pointed to an entirely new function for the entorhinal cortex and suggested that spatial signals in the hippocampus are derived from afferent areas rather than being computed in the structure itself. The hippocampus was thereby "released" from an exclusive spatial domain, and a primary role in the encoding and storage of associative memory, both spatial and non-spatial, seemed more likely.

Grid cells

The key element of the neural map in MEC is the "grid cell", a cell type that is different from any other functionally defined cell category in the nervous system. In our early recordings, we noticed that the majority of the cells in this brain area were active exclusively when the rat was at certain locations in the environment, and that these locations formed a regular pattern with a non-random spacing (Fyhn et al. 2004). In subsequent work, using recording enclosures that were large enough to capture the spatial structure of the firing fields of the entorhinal neurons, we observed that the activity peaks formed a highly periodic pattern (Hafting et al. 2005). For each cell, the set of active locations defined a regular array, or a grid, covering the entirety of the animal's environment, like the cross-points of graph paper rolled out over the surface of the test arena (Fig. 2A). The repeating unit, however, was not a square, like on the graph paper, but an equilateral triangle or, more precisely, a pair of oppositely oriented equilateral triangles. Grids of different cells were offset relative to each other (Fig. 2B), such that each position in the environment could be identified from the activity of a fairly limited number of adjacent cells. Using the collective activity of a small set of grid cells, we were able to show that it was possible to predict the animal's current location as it moved through space with a precision of a few

A

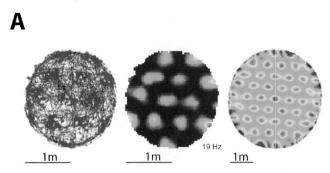

B

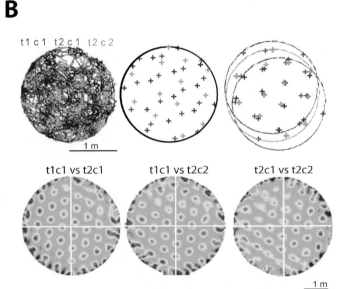

Fig. 2. Firing fields of grid cells in MEC. (**A**) A representative grid cell in layer II. *From left to right*: trajectory of the rat (*black*) with superimposed spike locations (*red*); color-coded rate map with peak rate indicated; and spatial autocorrelation of the rate map. The color scales are from *blue to red* (minimum to maximum rate, or autocorrelation from r = −1 to r = 1). (**B**) *Grid fields* of neighboring cells have different spatial phases, i.e., their grids are offset relative to each other. *Top row*: Combined maps of three simultaneously recorded grid cells (t1c1, t2c1, and t2c2; each shown in a different color). The left map shows the locations of all spikes on the animal's trajectory; the *middle map* shows the peak firing locations of each subfield; in the *right map*, the peaks of the three maps are offset to show the similarity in spacing and orientation. *Bottom row*: Spatial cross-correlation of the same three cells. Note that the peaks of the cross-correlograms are offset from the origin, indicating a phase shift in the spatial maps. Reproduced from Hafting et al. (2005)

centimetres (Fyhn et al. 2004). These observations point to the network of entorhinal grid cells as a possible neuronal coordinate system used by mammals during spatial navigation.

A key property of the newly discovered entorhinal spatial map is its apparently universal activation (Hafting et al. 2005). Unlike place cells in the hippocampus, the grid fields of entorhinal neurons are activated in a stereotypic manner across environments, regardless of the environment's particular landmarks, suggesting that the same neural map is applied wherever the animal is walking. Grids persist in darkness and appear almost instantaneously as an animal is introduced into a novel environment. Yet they are also controlled by extrinsic cues, as grid fields follow the landmarks when the walls of an environment are rotated. These observations suggest that, while landmarks are important for aligning grid representations with the particular landmarks of an environment, keeping representations stable from one occasion to the next, the dynamic computation of position is apparently based on the animal's own movement, using distance and direction of movement as the primary inputs, without reference to the external environment. The grid-cell map is thus likely to be part of the brain's mechanism for path integration (Mittelstaedt and Mittelstaedt 1980). Because of its grid-like nature, the map can potentially represent places not visited just as well as places that have been visited. The grid map is potentially infinite, allowing the inclusion of all places and potential places. The existence of a single neural map of the environment that can be applied anywhere is economical and avoids the enormous capacity problems that might arise if the brain were to store rules about spatial interrelations for every single spatial context that the subject encountered.

A striking feature of the entorhinal map is its regular organization, with grids of neighboring cells having a common spacing (distance between activity spots) and a common orientation (direction of the three axes). The spacing of the grid increases progressively as one gets deeper into the brain. The combination of grids at variable scales, within the entorhinal cortex or downstream in the hippocampus, provides an economical, high-resolution spatial coordinate system for navigation over a large space. If grid cells had a single, common scale, the population code would repeat itself at intervals corresponding to a single period of the grid, and the place code would be ambiguous. If activity is integrated across grids with different spacings and orientations, however, the cycle for repetition might be very large, enabling each position within this radius to be expressed by a unique pattern of collective activity in the grid network. In contrast to the topographic organization of spacing and orientation, the phase of the grid – the relative displacement of grid vertices in different cells – is randomly distributed. These organizational features are all mirrored downstream in the hippocampus, in which neighboring cells have different firing locations and properties (O'Keefe and Nadel 1978; Hirase et al. 2001; Redish et al. 2001), and the scale of the place fields increases along the septal-to-temporal axis of the brain area (Jung et al. 1994; Maurer et al. 2005; Kjelstrup et al. 2006).

Head-direction cells and conjunctive cells

The regular structure of the grid field and its relative independence of landmarks in the environment implicate the grid cell as part of a universal, path-integration-based spatial metric (Hafting et al. 2005; Fuhs and Touretzky 2006; McNaughton et al. 2006). A role for grid cells in path integration is also consistent with the inability of animals with entorhinal cortex lesions to calculate a return path to their home cage on the basis of self-motion cues (Parron and Save 2004). How the position vector is computed is not understood, however. To explore how grid cells interact with cells in other layers of the MEC, we recorded from each principal cell layer of MEC in rats that explored two-dimensional environments (Sargolini et al. 2006). Grid cells were found in all principal cell layers (Fig. 3A), but the proportion of cells with grid properties was layer-dependent. In layer II, most well-separated cells exhibited periodic firing, as in our earlier observations (Hafting et al. 2005). In the deeper layers, a subset of the neurons had clear grid-like firing fields, but the proportion of grid cells was smaller.

A

Grid cell

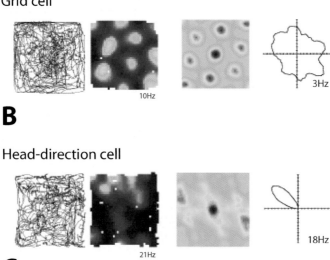

B

Head-direction cell

C

Conjunctive cell

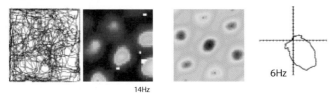

Fig. 3. Representation of position and direction in MEC neurons. *From left to right*: trajectory map with individual spike locations, color-coded rate map, autocorrelation matrix, and polar plot indicating firing rate as a function of head direction. Peak rate is indicated. Symbols for trajectory maps, rate maps and autocorrelograms are as in Fig. 2. (**A**) Representative grid cell from *layer II*. (**B**) Representative head-direction cell, recorded from *layer V*. (**C**) Representative grid x head-direction cell, recorded from *layer III*. Reproduced from Sargolini et al. (2006)

The scale of the grid increased from the dorsocaudal to the ventromedial end of the MEC, like in layer II, such that cells near the postrhinal border had the densest spacing and the smallest firing fields.

In all layers except layer II, grid cells intermingled with head-direction cells whose properties were similar to those of the head-direction cells recorded in dorsal presubiculum and anterior thalamus in previous studies (Taube et al. 1990a,b; Knierim et al. 1995). The firing rate of entorhinal head-direction cells increased from a very low background rate to peak rate of up to 40 Hz or more whenever the rat's head turned to a certain range of directions. Most head-direction cells had a sharp directional tuning that was similar over the entire apparatus and stable across trials (Fig. 3B). Their peak firing directions were widely distributed, and when several head-direction cells were recorded simultaneously, their preferred directions were usually uncorrelated (Fig. 4). Together with the distributed spatial phase of co-localized grid cells (Fig. 2B), and the

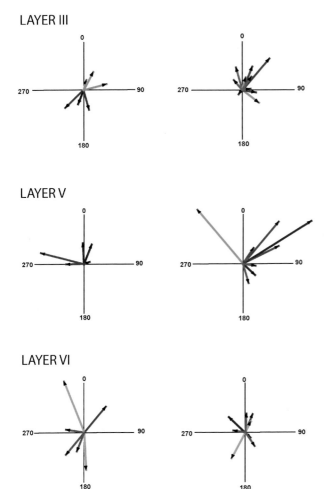

Fig. 4. Peak firing direction of co-localized MEC head-direction cells are widely distributed. *Each panel* shows a vector representation of the preferred firing direction (*angle*) and the peak rate (*length*) of simultaneously recorded head-direction cells, from layers III, V and VI (two recordings from *each layer*). Cells from the same tetrode are indicated with the same color. Reproduced from Sargolini et al. (2006)

co-localization of grid cells and head-direction cells (Fig. 5), the distributed organization of directional preferences may enable the representation of all possible positions, distances and directions within small areas of the MEC.

Grid cells and head-direction cells were found to form overlapping populations, where grid cells express variable degrees of directional modulation and head-direction cells express variable degrees of grid-like firing structure. Cells with dual firing properties were referred to as conjunctive cells (Fig. 3C). All conjunctive cells were also modulated by the animal's instantaneous speed of movement, enabling them to combine information about the origin of a trajectory with information about how far and in what direction the animal has moved at any given time. The presence of cells with conjunctive properties points to a possible mechanism for translation of the position representation in the grid-cell layer (Samsonovich and McNaughton 1997; Sargolini et al. 2006; McNaughton et al. 2006). One important clue is that conjunctive cells are located predominantly in layers III and V, where principal cells have extensive dendritic trees extending up to the pial surface (Lingenhohl and Finch 1991; Hamam et al. 2000; van Haeften et al. 2003; Kloosterman et al. 2003), which might enable them to be activated in a location-specific manner by grid cells in layer II. In addition, these principal neurons have extensive axonal projections back to the superficial layers, including, probably, the grid cells (van Haeften et al. 2003; Kloosterman et al. 2003). If these ascending projections are asymmetric, contacting cells with a slightly different spatial phase, the activation of conjunctive neurons by a specific combination of grid cells and directional input cells may result in translation of the position representation in the population of grid cells, i.e., a shift of activity to cells with a different spatial phase, in a manner consistent with the rat's motion. The direction of the translation may be determined by the head-direction input; the distance may be controlled by the speed modulation.

Cells with conjunctive grid x head-direction properties were encountered in all principal layers except for layer II. Together with the strong interlaminar connections of the entorhinal microcircuit, this finding implies that, despite the differential hippocampal and neocortical connections of superficial and deep layers of the MEC (Witter and Amaral 2004), the layers operate as an integrated unit, with significant

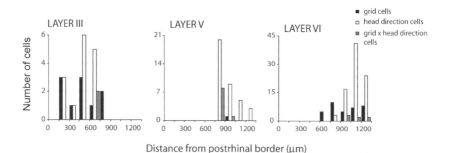

Fig. 5. Co-localization of grid cells, head-direction cells, and grid x head-direction cells in the MEC. Distribution of the number of cells in each category as a function of the recording depth (distance from the *postrhinal border*) in three representative rats (one rat for *each layer*). Reproduced from Sargolini et al. (2006)

interaction between grid cells, present in all principal cell layers, and head-direction cells and conjunctive cells, present in layers III to VI.

Conclusions

The discovery of grid cells has pointed to a possible neural basis for spatial representation and spatial memory in the brain. Thirty years of almost exclusive focus on spatial representation in the hippocampus has been broadened to include a wider cortical network, with the entorhinal cortex and associated parahippocampal areas as major contributors. As a prominent hub of this network, the medial entorhinal cortex is well suited to integrate information about distance, direction and position into an environmentally invariant, self-motion-based dynamic representation of current position during spatial navigation in mammals.

References

Barnes CA, McNaughton BL, Mizumori SJ, Leonard BW, Lin LH (1990) Comparison of spatial and temporal characteristics of neuronal activity in sequential stages of hippocampal processing. Prog Brain Res 83:287–300

Bostock E, Muller RU, Kubie JL (1991) Experience-dependent modifications of hippocampal place cell firing. Hippocampus 1:193–205

Burwell RD (2000) The parahippocampal region: corticocortical connectivity. Ann NY Acad Sci 911:25–42

Dolorfo CL, Amaral DG (1998) Entorhinal cortex of the rat: topographic organization of the cells of origin of the perforant path projection to the dentate gyrus. J Comp Neurol 398:25–48

Frank LM, Brown EN, Wilson M (2000) Trajectory encoding in the hippocampus and entorhinal cortex. Neuron 27:169–178

Fuhs MC, Touretzky DS (2006) A spin glass model of path integration in rat medial entorhinal cortex. J Neurosci 26:4266–4276

Fyhn M, Molden S, Witter MP, Moser EI, Moser M-B (2004) Spatial representation in the entorhinal cortex. Science 305:1258–1264

Hafting T, Fyhn M, Molden S, Moser M-B, Moser EI (2005) Microstructure of a spatial map in the entorhinal cortex. Nature 436:801–806

Hamam BN, Kennedy TE, Alonso A, Amaral DG (2000) Morphological and electrophysiological characteristics of layer V neurons of the rat medial entorhinal cortex. J Comp Neurol 418:457–472

Hirase H, Leinekugel X, Csicsvari J, Czurko A, Buzsaki G (2001) Behavior-dependent states of the hippocampal network affect functional clustering of neurons. J Neurosci 21:RC145

Hollup SA, Molden S, Donnett JG, Moser MB, Moser EI (2001) Accumulation of hippocampal place fields at the goal location in an annular watermaze task. J Neurosci 21:1635–1644

Jung MW, Wiener SI, McNaughton BL (1994) Comparison of spatial firing characteristics of units in dorsal and ventral hippocampus of the rat. J Neurosci 14:7347–7356

Kentros C, Hargreaves E, Hawkins RD, Kandel ER, Shapiro M, Muller RV (1998) Abolition of long-term stability of new hippocampal place cell maps by NMDA receptor blockade. Science 280:2121–2126

Kjelstrup KB, Solstad T, Brun VH, Fyhn M, Hafting T, Leutgeb S, Witter MP, Moser M-B, Moser EI (2006) Spatial scale expansion along the dorsal-to-ventral axis of hippocampal area CA3 in the rat. 5th Forum of European Neuroscience, Vienna, Austria

Kloosterman F, van Haeften T, Witter MP, Lopes da Silva FH (2003) Electrophysiological characterization of interlaminar entorhinal connections: an essential link for re-entrance in the hippocampal-entorhinal system. Eur J Neurosci 18:3037–3042

Knierim JJ, Kudrimoti HS, McNaughton BL (1995) Place cells, head-direction cells and the learning of landmark stability. J Neurosci 15:1648–1659

Leutgeb JK, Leutgeb S, Treves A, Meyer R, Barnes CA, McNaughton BL, Moser MB, Moser EI (2005a) Progressive transformation of hippocampal neuronal representations in "morphed" environments. Neuron 48:345–358

Leutgeb S, Leutgeb JK, Barnes CA, Moser EI, McNaughton BL, Moser MB (2005b) Independent codes for spatial and episodic memory in hippocampal neuronal ensembles. Science 309:619–623

Leutgeb S, Leutgeb JK, Moser MB, Moser EI (2005c) Place cells, spatial maps and the population code for memory. Curr Opin Neurobiol 15:738–746

Lever C, Wills T, Cacucci F, Burgess N, O'Keefe J (2002) Long-term plasticity in hippocampal place-cell representation of environmental geometry. Nature 416:90–94

Lingenhohl K, Finch DM (1991) Morphological characterization of rat entorhinal neurons in vivo: soma-dendritic structure and axonal domains. Exp Brain Res 84:57–74

Maguire EA, Frackowiak RS, Frith CD (1997) Recalling routes around London: activation of the right hippocampus in taxi drivers. J Neurosci 17:7103–7110

Maguire EA, Burgess N, Donnett JG, Frackowiak RS, Frith CD, O'Keefe J (1998) Knowing where and getting there: a human navigation network. Science 280:921–924

Markus EJ, Qin YL, Leonard B, Skaggs WE, McNaughton BL, Barnes CA (1995) Interactions between location and task affect the spatial and directional firing of hippocampal neurons. J Neurosci 15:7079–7094

Maurer AP, Vanrhoads SR, Sutherland GR, Lipa P, McNaughton BL (2005) Self-motion and the origin of differential spatial scaling along the septo-temporal axis of the hippocampus. Hippocampus 15:841–852

McNaughton BL, Battaglia FP, Jensen O, Moser EI, Moser M-B (2006) Path-integration and the neural basis of the "cognitive map". Nat Rev Neurosci 7:663–678

Mittelstaedt ML, Mittelstaedt H (1980) Homing by path integration in a mammal. Naturwissenschaften 67:566–567

Morris RMG, Garrud P, Rawlins JNP, O'Keefe JO (1982) Place navigation impaired in rats with hippocampal lesions. Nature 297:681–683

Nadel L (1991) The hippocampus and space revisited. Hippocampus 1:221–229

O'Keefe J, Dostrovsky J (1971) The hippocampus as a spatial map. Preliminary evidence from unit activity in the freely-moving rat. Brain Res 34:171–175

O'Keefe J, Nadel L (1978) The hippocampus as a cognitive map. Clarendon Press, Oxford

Olton DS, Walker JA, Gage FH (1978) Hippocampal connections and spatial discrimination. Brain Res 139:295–308

Parron C, Save E (2004) Evidence for entorhinal and parietal cortices involvement in path integration in the rat. Exp Brain Res 159:349–359

Quirk GJ, Muller RU, Kubie JL, Ranck JB Jr (1992) The positional firing properties of medial entorhinal neurons: description and comparison with hippocampal place cells. J Neurosci 12:1945–1963

Redish AD, Battaglia FP, Chawla MK, Ekstrom AD, Gerrard JL, Lipa P, Rosenzweig ES, Worley PF, Guzowski JF, McNaughton BL, Barnes CA (2001) Independence of firing correlates of anatomically proximate hippocampal pyramidal cells. J Neurosci 21:RC134

Samsonovich A, McNaughton BL (1997) Path integration and cognitive mapping in a continuous attractor neural network model. J Neurosci 17:272–275

Sargolini F, Fyhn M, Hafting T, McNaughton B, Witter MP, Moser M-B, Moser EI (2006) conjunctive representation of position, direction and velocity in the medial entorhinal cortex. Science 312:680–681

Scoville WB, Milner B (1957) Loss of recent memory after bilateral hippocampal lesions. J Neurol Neurosurg Psych 20:11–21

Squire LR, Stark CE, Clark RE (2004) The medial temporal lobe. Annu Rev Neurosci 27:279–306

Steffenach HA, Witter M, Moser MB, Moser EI (2005) Spatial memory in the rat requires the dorsolateral band of the entorhinal cortex. Neuron 45:301–313

Taube JS, Muller RU, Rank JB Jr (1990a) Head direction cells recorded from the postsubiculum in freely moving rats. I. Description and quantitative analysis. J Neurosci 10:420–435

Taube JS, Muller RU, Ranck JB Jr (1990b) Head-direction cells recorded from the postsubiculum in freely moving rats. II. Effects of environmental manipulations. J Neurosci 10:436–447

van Haeften T, Baks-te-Bulte TL, Goede PH, Wouterlood FG, Witter MP (2003) Morphological and numerical analysis of synaptic interactions between neurons in deep and superficial layers of the entorhinal cortex of the rat. Hippocampus 13:943–948

Wilson MA, McNaughton BL (1993) Dynamics of the hippocampal ensemble code for space. Science 261:1055–1058

Witter MP, Amaral DG (2004) The hippocampal formation. In: Paxinos G (ed) The rat nervous system.Edition 3. Academic Press, San Diego, pp 637–703

Witter MP, Groenewegen HJ, Lopes da Silva FH, Lohman AH (1989) Functional organization of the extrinsic and intrinsic circuitry of the parahippocampal region. Prog Neurobiol 33:161–253

Wood ER, Dudchenko PA, Eichenbaum H (1999) The global record of memory in hippocampal neuronal activity. Nature 397:613–616

Wood ER, Dudchenko PA, Robitsek RJ, Eichenbaum H (2000) Hippocampal neurons encode information about different types of memory episodes occurring in the same location. Neuron 27:623–633

The Prefrontal Cortex: Categories, Concepts, and Cognitive Control

Earl K. Miller

Summary

What controls your thoughts? How do you know how to act while dining in a restaurant? This is cognitive control, the ability to organize thought and action around goals. Results from our laboratory have shown that neurons in the prefrontal cortex and related brain areas have properties commensurate with a role in "executive" brain function. Perhaps most importantly, they transmit acquired knowledge. Here, I discuss how the prefrontal cortex and basal ganglia may help obtain our internal representations of rules and principles needed for goal-directed behavior, thereby providing the foundation for the complex behavior of primates, in whom the prefrontal cortex is most elaborate.

Introduction

Virtually all animals are capable of reacting to their immediate environment. Food is approached, predators avoided, etc. Many of these reactions are automatic processes and reflexive. If something is bearing down on us, we leap out of the way before we have even had to consciously form the thought. Automatic processes thus seem to depend on relatively straightforward, hardwired, relationships between the brain's input and output systems. In neural terms, it seems that they depend on strong, well-established, neural pathways just waiting to be triggered by the right sensory cue. That is, automatic processes are driven in a "bottom-up" fashion: they are determined largely by the nature of the sensory stimuli and whatever reaction they are most strongly wired to. But more advanced animals, such as primates, can also be proactive and goal-directed. We have the capacity to predict unseen goals and put together plans to achieve those that are within our reach. We can suppress reflexive reactions to the environment and urges in order to willfully direct behavior toward anticipated and desired future states. This is called cognitive control.

Learning and memory are at the essence of both classes of behavior. Automatic habits seem to be established by repeated co-activation of neural pathways, as if "ruts" were being carved into the brain. These habits can be simply triggered and fired off in a "ballistic" fashion with little variation; hence they require little internal oversight. In contrast, for truly sophisticated goal-directed behavior, simply recording and replaying

---

[1] The Picower Institute for Learning and Memory, RIKEN-MIT Neuroscience Research Center, and Department of Brain and Cognitive Sciences, Massachusetts Institute of Technology, 77 Massachusetts Avenue 46-6241, Cambridge, MA 02139, USA
ekmiller@mit.edu

Bontempi et al.
Memories: Molecules and Circuits
© Springer-Verlag Berlin Heidelberg 2007

experiences is not sufficient. Goal-relevant relationships need to be sorted out from spurious coincidences and, importantly, long-term goals require more than figuring out the world piecemeal, one situation at a time. Smart animals get the "big picture" of the jigsaw puzzle of their experiences, the common structure across a wide range of experiences. These generalized, abstracted representations offer an efficient way to deal with a complex world. They allow the navigation of many different situations with a minimal amount of storage and also allow us to deal with novelty. By extracting the essential elements from our experiences, we can generalize to future situations that share some elements but may, on the surface, appear very different. For example, consider the concept "camera". We do not have to learn anew about every camera we may encounter. Just knowing that the item is a camera communicates a great deal of knowledge about its parts, functions, and operations.

How and where such knowledge is acquired by the brain are key questions for understanding cognitive control. One cortical region, the prefrontal cortex (PFC), seems to play a central role. It is the brain area that reaches its greatest elaboration in the human brain and is thus thought to be the neural instantiation of the mental qualities that we think of as "intelligent". Here, I will discuss evidence for the role of the PFC in, and theories of, cognitive control, with an emphasis on how the PFC and related brain areas might acquire and represent the higher-order knowledge and abstract principles needed to orchestrate sophisticated goal-directed behavior.

The Prefrontal Cortex

The prefrontal cortex (PFC), the cortex at the very front of the brain, is an ideal place to look for neural correlates of abstract information. It occupies a far greater proportion of the human cerebral cortex than in other animals, suggesting that it might contribute to those cognitive capacities that separate humans from animals (Fuster 1995). At first glance, PFC damage has remarkably little overt effect; patients can perceive and move, there is little impairment in their memory, and they can appear remarkably normal in casual conversation. Despite the superficial appearance of normality, PFC damage seems to devastate a person's life. They have difficulty in sustaining attention and keeping "on task", and they seem to act on whims and impulses without regard to future consequences. This pattern of high-level deficits coupled with a sparing of lower-level, basic, functions has been called a "dysexecutive syndrome" (Baddeley and Della Sala 1996) and "goal neglect" (Duncan et al. 1996).

Indeed, the anatomy of the PFC suggests that it is well suited for a role as the brain's "executive". It can synthesize information from a wide range of brain systems and exert control over behavior (Nauta 1971; Barbas and Pandya 1991). The collection of cortical areas that comprise the PFC has interconnections with brain areas processing external information (with all sensory systems and with cortical and subcortical motor system structures) as well as internal information (limbic and midbrain structures involved in affect, memory, and reward). Correspondingly, its neurons are highly multimodal and encode many different types of information from all stages of the perception-action cycle (Fuster 1995). They are activated by stimuli from all sensory modalities, before and during a variety of actions, during memory for past events, and in anticipation of expected events and behavioral consequences, and they are modulated by internal factors such as motivational and attentional state (for review, see Miller and Cohen 2001). Because of its highly multimodal nature and its apparent role in higher mental life, the PFC seemed like an ideal place to begin our search for neural correlates of the abstract information needed for intelligent behavior.

Neural Correlates of Categories and Concepts

Categorical representations do not faithfully track exact sensory input. They provide useful groupings and divisions not present in the external world and thus are the building blocks of the high-level knowledge needed for sophisticated, goal-directed behavior. Consider a simple example. Crickets sharply divide a range of pure tones into "mate" versus "bat" (a predator; Wyttenbach et al. 1996). Even though the input varies along a continuum, behavior does not. At the low end of the range, crickets approach the sound equally, but then at 16 kHz, their behavior suddenly flips to avoidance and remains equivalent across another wide range. This behavior allows the crickets to maximize reproduction while minimizing disaster. Another example is humans' perception of the facial expressions of emotion (Beale and Keil 1995), which also flips at a discrete point. Thus, the representation of perceptual categories must involve something beyond the sort of neural tuning that encodes physical attributes, i.e., gradual changes in neural activity as attributes gradually change (e.g., shape, orientation, direction). Perceptual categories instead have sharp boundaries (not gradual transitions), and members of the same category are treated as similar even though their physical appearance may vary widely. This is the most fundamental form of categorization. More conceptual categories can have "fuzzy" boundaries and are likely to rely on different types of representation.

Shape-based categories

By learning to take multiple dimensions into account to make sharp distinctions and groupings, advanced animals such as primates can acquire relatively complex perceptual categories, such as "animal" (Roberts and Mazmanian 1988), "food" (Fabre-Thorpe et al. 1998), and "fish" (Vogels 1999a). The search for neural correlates of such high-level, shape-based categories has naturally focused on brain regions at the final stages of visual form processing, the inferior temporal cortex (ITC: Desimone et al. 1984; Logothetis and Sheinberg 1996; Tanaka 1996) and the PFC, which receives highly processed visual information and orchestrates voluntary, goal-directed behaviors (Fuster 1995; Miller and Cohen 2001).

The ITC has long been known to contain neurons with a high degree of selectivity for complex objects. Neurons with tuning properties suggestive of categories have been identified since the seminal work of Gross and colleagues (Desimone et al. 1984). They described a small population of "face cells" that were strongly activated by the gestalt of a face but not by individual features or other stimuli. These neurons tend to be highly localized and clustered in both the monkey and human temporal cortex, as are neurons selectively activated by pictures of familiar places (Kanwisher et al. 1997; Epstein and Kanwisher 1998; Tsao et al. 2006). Vogels (1999b) reported a small number of inferior temporal neurons in trained monkeys that were specifically activated by "trees" or "fish" and showed relatively little differentiation between diverse examples of those categories. Kreiman et al. (2000) recorded from the hippocampus, amygdala, and adjacent cortex of human epileptic patients and found a few neurons that were highly selective for diverse pictures of familiar concepts like "Bill Clinton".

This use of natural images has identified neurons that ultimately contribute to categorical representations, but unless neurons are tested for the identifying characteristics of perceptual categories (sharp boundaries, relative equivalence within a category), it is possible that their activity reflects physical similarity rather than category membership per se; "trees", after all, look more like one another than like other stimuli. This is not to say that neurons with such complex properties are not critical contributors to categories. The question is whether they are the *only* contributors. Are category representations the sum of neurons sensitive to the physical appearance of its members or are there neurons that favor information about category groupings over information about the individuals and explicitly represent category per se? That is, are there neurons whose activity mirrors the across-category distinctions and within-category generalizations seen on the behavioral level?

To test for neural correlates of perceptual categories, we trained monkeys to categorize computer-generated stimuli into two categories, "cats" and "dogs" (Freedman et al. 2001; Fig. 1). A novel 3D morphing system was used to create a large set of parametric blends of six prototype images (three species of cats and three breeds of dogs; Beymer and Poggio 1996; Shelton 2000). By blending different amounts of cat and dog, we could smoothly vary shape and precisely define the boundary between the categories (> 50% of a given type). As a result, stimuli that were close to but on opposite sides of the boundary were physically similar, whereas stimuli that belonged to the same category could be physically dissimilar (e.g. the "cheetah" and "housecat").

Nearly one third of responsive neurons were category-selective in that they exhibited an overall difference in activity to cats versus dogs. Similar numbers gave stronger activity to cats as dogs. Figure 2 shows a single neuron that exhibited the hallmarks of perceptual categorization. It exhibited greater activity to dogs than cats and responded similarly to samples from the same category regardless of their degree of dogness or catness. This explicit reflection of the cat and dog categories in single neuron activity, just like lower-level attributes such as shape or color, may account for the ability of primates to rapidly and effortlessly categorize the world around us.

As our monkeys had no experience with cats or dogs prior to training, it seemed likely that these effects resulted from training. To test this, we defined two new category boundaries that were orthogonal to the original boundary (see Fig. 1; Freedman et al. 2001, 2002). Following training, we found that the original categories were no longer reflected in activity; instead, the three new categories were. A category index revealed no significant effect of the two old, now-irrelevant categories, either across the population of stimulus-selective neurons or across all recorded neurons. However, when the category index was computed using the three new (relevant) category boundaries, a significant category effect was observed across these populations, indicating that neural activity shifted to reflect the newly learned categories.

The next question concerned the contribution of visual cortex, namely the ITC. The ITC has long been thought to be the final stage in a visual cortex pathway needed for object recognition. Damage to it causes deficits in visual discrimination and learning (Gross 1973; Mishkin 1982) and category-specific agnosias (e.g., for faces) in humans

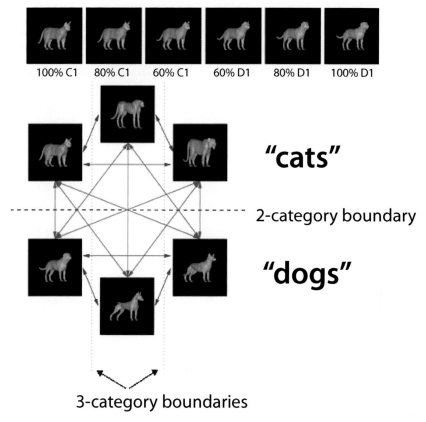

100% C1   80% C1   60% C1   60% D1   80% D1   100% D1

"cats"

2-category boundary

"dogs"

3-category boundaries

Fig. 1. "Cat" and "dog" prototypes and morph lines. Stimulus space was divided by one category boundary for cats vs dogs and by two boundaries that created three arbitrary categories. Thousands of unique morphs were created from the prototypes. *Red lines* are between-category and *blue lines* within-category *morph lines* for the cats and dogs. Examples of stimuli along the C1-D1 *morph line* are shown above

(Gainotti 2000). It might be that the category information in the PFC was retrieved from long-term storage in the ITC for its immediate use on the task. Interactions between the PFC and ITC underlie the storage and/or recall of visual memories and associations (Rainer et al. 1999; Tomita et al. 1999).

To determine whether PFC category effects were simply "inherited" from the ITC, we recorded from the ITC during performance of our category-matching task. This experiment yielded a striking difference. There was virtually no category effect across the ITC population and no examples of neurons whose activity showed the sharp across-distinction/within-category generalization that is the behavioral signature of categorization. Rather, ITC neurons were sensitive to the physical appearance of the individual stimuli; its neurons did not "throw away" information about individuals in favor of the category groupings, like PFC neurons did. This finding suggests that learned visual categories are abstracted at the level of the PFC, not in visual cortex.

Fig. 2. The shape category effect on the behavioral level (*top*) and in a single neuron. Note that both show a sharp transition at the category boundary. The *six colored lines* show the neuron's average activity to stimuli at different morph levels. The *colored tiles* show activity to individual stimuli along the *nine morph lines* that crossed the category boundary. The *bottom* shows recording sites where shape category neurons were found

But how are these categories abstracted? Some theories posit that we build up some form of average prototypes of each category, others a "list" of defining features. Indeed, we found in separate psychophysical experiments that each monkey learned a different combination of features to categorize the stimuli (Freedman et al. 2002). Insight into this process comes from a study by Sigala and Logothetis (2002). They found that stimulus features relevant for category judgments were enhanced in inferior temporal activity relative to irrelevant features. This finding suggests a process that weights features according to their relevance for category membership, supporting more feature-based models of categorization. Further evidence comes from Baker et al. (2002), who recently showed that training monkeys that certain features of complex objects "go together" results in ITC neural selectivity for those feature combinations. These studies all suggest that, while the ITC may emphasize and knit together behaviorally relevant features, the PFC is where this information converges to produce explicit representation of learned categories that largely discards information about individual category members.

Numerosity

The shape category effects in the lateral PFC naturally raised the question of whether it is specialized for this type of categorization or whether its role in categorization is broader. Small number judgments are markedly different from shape categories. Behavioral experiments and modelling work suggest that small number estimations do not rely on sharp boundaries like shape categories but overlapping filters tuned to different numbers (Dehaene and Changeux 1993; Dehaene et al. 1999). At the same time, numerosity is very abstract; "two" can refer to a pair of anything. Finally and importantly, unlike the learned cat and dog shape categories, small number estimation is *innate*; many animals can make number judgments without any training (Brannon and Terrace 1998; Brannon et al. 2001), as can preverbal human infants (Hauser and Hauser 2000; Xu and Spelke 2000; Hauser et al. 2002). Of course, animals cannot derive numerical information by verbally and serially "counting" items (like humans); rather they encode numerosities in a non-verbal, analog magnitude format (as do infants: Xu and Spelke 2000). Such approximate representations of numerical values may be regarded as biological precursors of adult humans' counting abilities.

Human imaging studies indicate that the frontal cortex and the posterior parietal cortex (Dehaene et al. 1999) are involved in numerical judgments. So, studies in monkeys have focused on those regions. Sawamura et al. (2002) trained monkeys to alternate between five arm movements of one type and five of another. They found neurons in a somatosensory-responsive region of the superior parietal lobule (SPL) that kept track of the movement number. By contrast, they found that relatively few such neurons were found in the lateral PFC, where we found our shape category neurons.

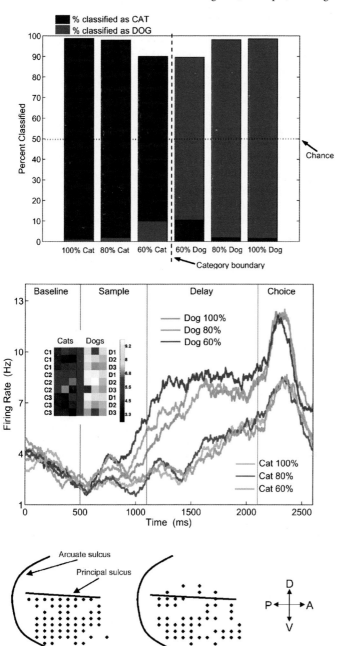

We trained monkeys to judge the number of items (1–5) in visual displays (Nieder et al. 2002). We trained monkeys to generalize by having them judge the number of items across a large number of very different looking displays (Fig. 3). We found ample number-tuned neurons (again, about 1/3 of the population) in the same portion of the lateral PFC where we had observed the shape category neurons (Fig. 4). One possibility for the difference between our study and that of Sawamura et al. may be modality (touch vs. vision), but another may be the level of abstraction. Most movement-number representations found by Sawamura and colleagues (85%) were not abstract: number-selective activity depended on whether the monkey's movement was "push" or "turn". By contrast, the visual number representations we found in the lateral PFC were abstract and generalized; activity was unaffected by the physical appearance of the displays.

We also found numerosity representations in posterior cortex: there was a small cluster of visual numerosity neurons in the intraparietal sulcus, as predicted from human imaging studies. However, numerosity neurons were far fewer in number than what was seen in the lateral PFC (and there were no numerosity effects in the inferior temporal cortex; Nieder and Miller 2004). This finding suggests a fronto-parietal network for numerosity in monkeys and establishes homologies between the monkey and human brain. Further, it indicates that training results in large fractions of PFC neurons conveying category information even when the category judgments involve special, innate abstractions like small numbers. It may be that innate categories have explicit representation in posterior cortical areas like numerosity in the parietal cortex (or faces in the inferior temporal cortex), but then large populations of PFC neurons become tuned to them as a result of training. By contrast, if there are no innate representations, the categories are abstracted de novo in the PFC (like our cat and dog categories).

Rules

Meaningful groupings of information do not only occur in the sensory domain. We can lump together sets of events and actions into general guidelines, principles, or rules for behavior. Consider restaurants. We have knowledge of generic rules. such as "wait to be seated", "order", and "pay the bill", that are long decoupled from the specific experiences in which they were learned. We then have a notion of what is expected of us the first time we walk into a new restaurant. As learning and applying abstracted rules and principles are central to complex, goal-directed behavior, a number of investigators have recently argued that their acquisition and representation may be a critical PFC function and their lack could explain the "goal neglect" that follows prefrontal damage (Houk and Wise 1995; Fuster 2000; Fuster et al. 2000; Miller and Cohen 2001; Wood et al. 2005). Neural correlates of rules like shape categories and numerosity have been shown to be abundant in the PFC of both monkeys and rodents (Asaad et al. 1998, 2000; Hoshi et al. 1998; White and Wise 1999; Wallis et al. 2000; Schoenbaum and Setlow 2001), suggesting that the visual categories described above are a sensory-based manifestation of a general prefrontal function in abstracting information across many domains.

When rules involve familiar cues and responses (such as "stop at red"), they are "concrete" and can be represented as a set of specific Stimulus-Response (S-R) associations. With varied experience, rules can be abstracted beyond such specificity. Wallis

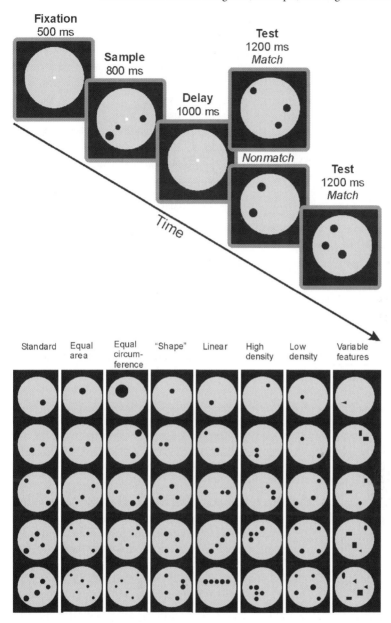

Fig. 3. Number-matching task. Monkeys were trained to match the number of items in a display across randomly generated changes in item size and position. After training, the monkeys and neurons generalized across disparate displays that controlled for changes in low-level attributes

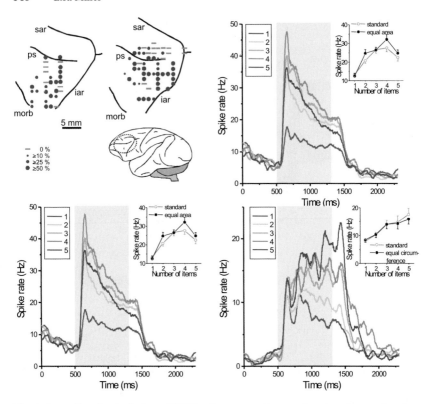

Fig. 4. Recording sites in the lateral PFC for the numerosity experiment, and examples of neurons that are tuned for number and generalize across changes in the physical appearance of the displays. The recording sites show the percentage of neurons at a given recording site that exhibited numerosity tuning. We found these neurons in virtually the same regions of the lateral PFC where we found the cat and dog neurons

et al. trained monkeys to do just that (Wallis et al. 2001; Wallis and Miller 2003; Muhammad et al. 2006). Monkeys alternated between applying either a "same" or "different" rule to pairs of pictures. After training with a wide range of pictures, they could apply these rules successfully upon seeing a novel picture for the first time (and therefore had no prior S-R association). This finding means the monkeys had learned to apply generalized principles. About 40% of neurons in the PFC reflected which rule was currently being used (Fig. 5). This neural activity was also generalized: it was independent of which specific cue signalled the rule, not linked to the behavioral response, and unaffected by which pictures the monkeys were judging. PFC engagement by abstract rules has also been reported in human imaging studies (Bunge et al. 2003).

We also tested the inferior temporal cortex because the ITC is presumably involved in visual analysis of the pictures used in the task (and, as mentioned above, is directly connected with the PFC). In sharp contrast, there was virtually no abstract rule information in the ITC (Muhammad et al. 2006). This finding parallels our work on shape

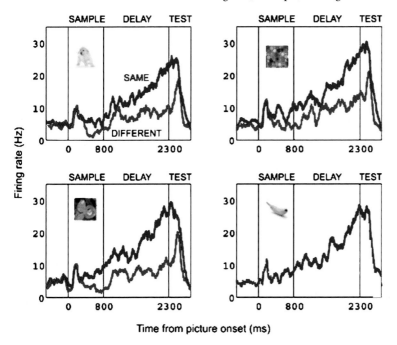

Fig. 5. Illustration of the firing rate of a single PFC neuron across time when the monkey is remembering one of four different pictures, and which of two abstract rules it should follow: "same" or "different". The sample picture is shown to the monkey for 800 ms. There is then a 1500-ms delay before the test picture appears. At this point, the monkey must decide whether the test picture is the same as the sample picture or different. Typically, the firing rate of PFC neurons does not encode the identity of the picture but encodes which abstract rule the monkey was following. The neuron illustrated here showed higher activity when the monkey was remembering the "same" rule compared to the "different" rule. The four histograms show the neuron's activity when the monkey was holding four different pictures in memory. Note that rule activity is the same regardless of which picture was remembered

categories and thus adds further support to the notion that the learning of high-level, goal-relevant abstraction is an important frontal cortex function.

Slow versus fast learning and abstraction

At first blush, it might seem that the greatest benefit would come from learning that proceeds as quickly as possible. Fast learning has obvious advantages: one can learn to get to resources and avoid obstacles faster and better than competitors. But fast learning comes at a cost; it does not allow the benefits that come from generalizing over multiple experiences, so by necessity it tends to be specific. Extending learning across multiple episodes allows organisms to pick up on the regularities of predictive relationships and allows detection of common structure across many different experiences, the regularities and commonalities that form abstractions, general principles, concepts, and symbolisms that are the medium of the sophisticated thought and the "big pictures" needed for truly long-term goals. Indeed, this is key to proactive thought and action. Generalizing among many past experiences naturally endows the ability to generalize to the future, to imagine possibilities that we have not yet experienced, but would like to. So, how does the brain balance the obvious pressure to learn as quickly as possible with the advantages of slower learning in forming abstractions? Some recent evidence suggests that this process may take place in interactions between the PFC and the basal ganglia.

The basal ganglia (BG) is a collection of subcortical nuclei. Cortical inputs arrive largely via the striatum (which includes both the caudate and the putamen), are processed through the globus pallidus (GP), the subthalamic nucleus (STN), and the substantia nigra (SN), and are then directed back into the cortex via the thalamus. The frontal cortex receives the largest portion of BG outputs, suggesting some form of close collaboration between them (Middleton and Strick 1994, 2000, 2002).

We recently compared activity in the lateral PFC and BG (specifically the caudate nucleus, which receives direct inputs from the PFC) as monkeys learned to associate a visual cue and a directional eye movement (Pasupathy and Miller 2005). Over the few tens of trials it took for monkeys to learn these associations, neural activity in the striatum showed rapid, almost bi-stable, changes compared to a much slower trend in the PFC. Interestingly, however, the slower, more executive PFC seemed to be the final arbiter of behavior; the monkeys' improvement in selecting the correct response more closely matched PFC changes.

This finding fits with observations of a striatal infrastructure ideal for rapid, supervised (reward-based) learning (Wilson and Kawaguchi 1996; Reynolds et al. 2001). It also supports hypotheses that learning in the frontal cortex is "trained" by the BG (Houk and Wise 1995; O'Reilly and Munakata 2000). Dopaminergic reward-prediction error signals from the midbrain (Schultz and Dickinson 2000; McClure et al. 2003) may allow rapid formation of reward-relevant associations in the striatum (Bar-Gad et al. 2003), which then trains slower, more graded, Hebbian mechanisms in the PFC via output nucleii of the BG and thalamus (O'Reilly and Munakata 2000; O'Reilly and Frank 2006). If so, this may mean that the BG is more specialized for the rapid learning of specific information needed to produce a desirable outcome. By contrast, the slower changes in the PFC may have allowed it to accumulate more evidence and more slowly and judiciously arrive at the correct answer, as well as detect the generalities across more experiences needed to form abstractions.

The faster changes and stronger effects of the specific cue-response associations in the striatum of the BG than in the PFC seem consistent with this hypothesis. But if it is true then, by contrast, we might expect to find that abstract rules and general principles are more strongly encoded in the PFC. A recent experiment by Muhammad et al. showed just that (2006). Building on the work of Wallis et al. (2001), we trained monkeys to apply the abstract rules "same" and "different" to pairs of pictures. If the "same" rule was in effect, monkeys responded if the pictures were identical, whereas if the "different" rule was in effect, monkeys responded if the pictures were different. The rules were abstract since the monkeys were able to apply the rules to novel stimuli for which there could be no pre-existing S-R association. This is the definition of an abstract rule. Muhammad et al. recorded neural activity from the same PFC and striatal regions as Pasupathy and Miller (2005) and found that, in contrast to the specific-cue response associations, the abstract rules were reflected more strongly in PFC activity (more neurons with effects and larger effects) than in BG activity, the opposite of what Pasupathy and Miller (2005) reported for the specific cue-response associations.

Finally, another important aspect of the connections between the frontal cortex and BG is that they form "closed" anatomical loops. Anatomical tracing techniques have suggested that functionally similar cortical areas project into the same striosome (Yeterian and Van Hoesen 1978; Van Hoesen et al. 1981; Flaherty and Graybiel 1991). For example, both sensory and motor areas relating to the arm seem to preferentially innervate the same striosome. The BG maintains a degree of topographical separation in different "channels" throughout its nuclei, ensuring that the output via the thalamus is largely to the same cortical areas that gave rise to the initial inputs to the BG (Selemon and Goldman-Rakic 1985; Alexander et al. 1986; Parthasarathy et al. 1992; Hoover and Strick 1993; Kelly and Strick 2004).

This structure suggests a recursive system where the results from one iteration can be fed back through the loop for further processing. Such a system is ideal for bootstrapping: repeated iterations can link in more and more information, building itself into ever-increasing elaboration and sophistication. This process may allow the bootstrapping of neural representations to increasing complexity and, with the slower learning in the PFC, greater and greater abstractions. It may also underlie a hallmark of human intelligence: it is easiest for us to understand new concepts if they can be grounded in familiar ones. We learn to multiply through serial addition and we understand quantum mechanics by constructing analogies to waves and particles. Interactions between the BG and the prefrontal cortex may support this type of cognitive bootstrapping. As more complex and generalized representations are learned in the prefrontal cortex, they are passed down through the BG for further expansion.

The prefrontal cortex and BG: goal-directed learning and cognitive control

Here is the general idea. Goal-directed thought and actions are learned via dopaminergic-gated plasticity in frontal cortex-BG loops. Fast learning mechanisms in the BG (specifically the striatum) are more specialized for the detection and storage of specific experiences that can and do lead to reward (i.e., activation of the midbrain dopamine signals). The output of the BG trains slower learning mechanisms in the frontal cortex. The slower cortical learning is not only less error-prone, but it also allows the frontal cortex to build up abstract, generalized, representations that reflect the regularities across many different experiences. Recursive iterations of these loops allow bootstrapping of ever-more complex, and ever-more predictive, rules and greater abstractions.

In this view, the more-primitive BG is the "engine" driving goal-directed learning. The frontal cortex, and the PFC more specifically, is an "add on"; it evolved to add greater sophistication to the learning. Both the PFC and BG can drive behavior (they both send projections to cortical and subcortical motor system structures). However, the PFC may have greater leverage because it has widespread cortical connections and thus can influence cortical processing at many levels. By virtue of its feedback projections to sensory cortex, for example, the PFC can play a direct role in filtering out potentially distracting sensory information (i.e., attention). Thus, the PFC and BG may work together as different parts of a frontal lobe system for goal-directed learning, but because the PFC has the big picture and can exert a top-down influence on much of the cortex, it may be the executive component of this system.

It is important to note that this review was not meant to be an exhaustive survey of PFC and BG function, neither was it meant to imply that these are the only functions of these structures. For example, work by Kesner and colleagues suggest that the BG may select behavioral responses based, in part, on inputs from the PFC (Kesner and Rogers 2004). Models by Bullock and Grossberg suggest that the BG has a gating function, allowing or disallowing expression of certain neural representations in the PFC (and other brain areas; Brown et al. 2004). Graybiel and colleagues emphasized the role of the BG in the formation of habits (Graybiel 1998). We also point the reader to the work of Robbins, Arnsten, and colleagues on the role of ascending neurotransmitter systems in, and interactions between, the PFC and BG in mediating cognitive flexibility (Arnsten and Robbins 2002; Robbins 2005). None of these findings is inconsistent with what we propose here. This review is only one slice through the multivariate functions of these structures, one that emphasizes their role in acquiring the knowledge needed for sophisticated goal-directed behavior.

But how is this acquired knowledge used to coordinate the brain-wide processing needed for cognitive control? Miller and Cohen (2001) suggested that the cardinal PFC function is to acquire and actively maintain patterns of activity that represent goals and the means to achieve them (rules) *and* the cortical pathways needed to perform the task ("maps", hence "rulemaps"; Fig. 6). Under this model, activation of a PFC rulemap sets up bias signals that propagate throughout much of the rest of the cortex, affecting sensory systems as well as systems responsible for response execution, memory retrieval, emotional evaluation, etc. The aggregate effect is to guide the flow of neural activity along pathways that establish the proper mappings between

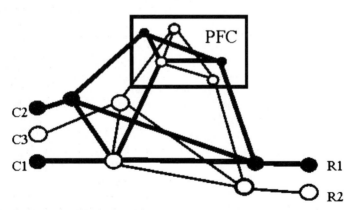

Fig. 6. In the Miller and Cohen model, a task model is formed in the PFC when reward signals link together neurons activated by the events that lead to reward. A subset of cues can then activate the entire representation. Bias signals resulting from the maintenance of this pattern guide the flow of activity along neural pathways that establish the task-relevant mappings between representations of inputs (C1, C2, C3), internal states, and outputs (R1, R2) in posterior cortex. *Thick lines* and *filled circles* denote active processing lines

inputs, internal states, and outputs to best perform the task. Establishing the proper mapping is especially important whenever stimuli are ambiguous (i.e., they activate more than one input representation), or when multiple responses are possible and the task-appropriate response must compete with stronger, more habitual, alternatives. In short, the task information is acquired by the PFC-BG loops, which when actively represented in the PFC, provides support to related information in posterior brain systems, effectively acting as a global attentional controller, or traffic cop, if you will. This endows the hallmarks of cognitive control. It provides *goal-direction* because there is an explicit representation of goals and means. It allows the brain to *override habits* by the PFC providing top-down signals that bias the flow of activity in the cortex away from strongly established (habit) pathways and toward weaker ones that match the PFC goal-related representation. Finally, it provides *flexibility* because, if you need to change your goal or strategy, you do not have to rewire your whole brain. If PFC activity patterns dynamically establish neural pathways throughout the cortex, then changing your mind is as easy and fast as changing the PFC activity pattern. Change the PFC pattern and the rest of the cortex falls in line.

References

Alexander GE, DeLong MR, Strick PL (1986) Parallel organization of functionally segregated circuits linking basal ganglia and cortex. Annu Rev Neurosci 9:357–381

Arnsten AFT, Robbins TW (2002) Neurochemical modulation of prefrontal cortical function. In: Stuss DT, Knight RT (eds) Principles of frontal lobe function. New York: Oxford University Press, pp 51–84

Asaad WF, Rainer G, Miller EK (1998) Task-related topography of neural activity in the primate prefrontal (PF) cortex. Soc Neurosci Abs 24:1425

Asaad WF, Rainer G, Miller EK (2000) Task-specific activity in the primate prefrontal cortex. J Neurophysiol 84:451–459

Baddeley A, Della Sala S (1996) Working memory and executive control. Phil Trans Roy Soc London B:Biol Sci 351:1397–1403

Baker CI, Behrmann M, Olson CR (2002) Impact of learning on representation of parts and wholes in monkey inferotemporal cortex. Nature Neurosci 5:1210–1216

Barbas H, Pandya D (1991) Patterns of connections of the prefrontal cortex in the rhesus monkey associated with cortical architecture. In: Levin HS, Eisenberg HM, Benton AL (eds) Frontal lobe function and dysfunction. New York: Oxford Univ. Press, pp 35–58

Bar-Gad I, Morris G, Bergman H (2003) Information processing, dimensionality reduction and reinforcement learning in the basal ganglia. Prog Neurobiol 71:439–473

Beale JM, Keil FC (1995) Categorical effects in the perception of faces. Cognition 57:217–239

Beymer D, Poggio T (1996) Image representations for visual learning. Science 272:1905–1909

Brannon EM, Terrace HS (1998) Ordering of the numerosities 1 to 9 by monkeys. Science 282:746–749

Brannon EM, Wusthoff CJ, Gallistel CR, Gibbon J (2001) Numerical subtraction in the pigeon: evidence for a linear subjective number scale. Psychol Sci 12:238–243

Brown JW, Bullock D, Grossberg S (2004) How laminar frontal cortex and basal ganglia circuits interact to control planned and reactive saccades. Neural Network 17:471–510

Bunge SA, Kahn I, Wallis JD, Miller EK, Wagner AD (2003) Neural circuits subserving the retrieval and maintenance of abstract rules. J Neurophysiol 90:3419–3428

Dehaene S, Changeux JP (1993) Development of elementary numerical abilities: A neural model. J Cogn Neurosci 5:390–407

Dehaene S, Spelke E, Pinel P, Stanescu R, Tsivkin S (1999) Sources of mathematical thinking: behavioral and brain-imaging evidence. Science 284:970–974

Desimone R, Albright TD, Gross CG, Bruce C (1984) Stimulus-selective properties of inferior temporal neurons in the macaque. J Neurosci 4:2051–2062

Duncan J, Emslie H, Williams P, Johnson R, Freer C (1996) Intelligence and the frontal lobe: The organization of goal-directed behavior. Cogn Psychol 30:257–303

Epstein R, Kanwisher N (1998) A cortical representation of the local visual environment. Nature 392:598–601

Fabre-Thorpe M, Richard G, Thorpe SJ (1998) Rapid categorization of natural images by rhesus monkeys. Neuroreport 9:303–308

Flaherty AW, Graybiel AM (1991) Corticostriatal transformations in the primate somatosensory system. Projections from physiologically mapped body-part representations. J Neurophysiol 66:1249–1263

Freedman DJ, Riesenhuber M, Poggio T, Miller EK (2001) Categorical representation of visual stimuli in the primate prefrontal cortex. Science 291:312–316

Freedman DJ, Riesenhuber M, Poggio T, Miller EK (2002) Visual categorization and the primate prefrontal cortex: Neurophysiology and behavior. J Neurophysiol 88:914–928

Fuster JM (1995) Memory in the cerebral cortex. Cambridge, MA: MIT Press

Fuster JM (2000) Executive frontal functions. Exp Brain Res 133:66–70

Fuster JM, Bodner M, Kroger JK (2000) Cross-modal and cross-temporal association in neurons of frontal cortex. Nature 405:347–351

Gainotti G (2000) What the locus of brain lesion tells us about the nature of the cognitive defect underlying category-specific disorders: a review. Cortex 36:539–559

Graybiel AM (1998) The basal ganglia and chunking of action repertoires. Neurobiol Learn Mem 70:119–136

Gross CG (1973) Visual functions of inferotemporal cortex. In: Jung R (ed) Handbook of sensory physiology. Berlin: Springer-Verlag 7:451–482

Hauser MC, S; Hauser, LB (2000) Spontaneous number representation in semi-free-ranging rhesus monkeys. Proc R Soc Lond B Biol Sci 267:829–833

Hauser MD, Dehaene S, Dehaene-Lambertz G, Patalano AL (2002) Spontaneous number discrimination of multi-format auditory stimuli in cotton-top tamarins (Saguinus oedipus). Cognition 86:B23–32

Hoover JE, Strick PL (1993) Multiple output channels in the basal ganglia. Science 259:819–821

Hoshi E, Shima K, Tanji J (1998) Task-dependent selectivity of movement-related neuronal activity in the primate prefrontal cortex. J Neurophysiol 80:3392–3397

Houk JC, Wise SP (1995) Distributed modular architectures linking basal ganglia, cerebellum, and cerebral cortex: their role in planning and controlling action. Cereb Cortex 5:95–110

Kanwisher N, McDermott J, Chun MM (1997) The fusiform face area: A module in human extrastriate cortex specialized for face perception. J Neurosci 17:4302–4311

Kelly RM, Strick PL (2004) Macro-architecture of basal ganglia loops with the cerebral cortex: use of rabies virus to reveal multisynaptic circuits. Prog Brain Res 143:449–459

Kesner RP, Rogers J (2004) An analysis of independence and interactions of brain substrates that subserve multiple attributes, memory systems, and underlying processes. Neurobiol Learn Mem 82:199–215

Kreiman G, Koch C, Fried I (2000) Category-specific visual responses of single neurons in the human medial temporal lobe. Nature Neurosci 3:946–953

Logothetis NK, Sheinberg DL (1996) Visual object recognition. Ann Rev Neurosci 19:577–621

McClure SM, Berns GS, Montague PR (2003) Temporal prediction errors in a passive learning task activate human striatum. Neuron 38:339–346

Middleton FA, Strick PL (1994) Anatomical evidence for cerebellar and basal ganglia involvement in higher cognitive function. Science 266:458–461

Middleton FA, Strick PL (2000) Basal ganglia and cerebellar loops: motor and cognitive circuits. Brain Res Rev 31:236–250

Middleton FA, Strick PL (2002) Basal-ganglia projections to the prefrontal cortex of the primate. Cereb Cortex 12:926–935

Miller EK, Cohen JD (2001) An integrative theory of prefrontal function. Ann Rev Neurosci 24:167–202

Mishkin M (1982) A memory system in the monkey. Philos Trans R Soc Lond B Biol Sci. 298:83–95

Muhammad R, Wallis JD, Miller EK (2006) A comparison of abstract rules in the prefrontal cortex, premotor cortex, the inferior temporal cortex and the striatum. J Cogn Neurosci, 6:974–989

Nauta WJH (1971) The problem of the frontal lobe: A reinterpretation. J Psychiatr Res 8:167–187

Nieder A, Miller EK (2004) A parieto-frontal network for visual numerical information in the monkey. Proc Natl Acad Sci USA 101:7457–7462.

Nieder A, Freedman DJ, Miller EK (2002) Representation of the quantity of visual items in the primate prefrontal cortex. Science 297:1708–1711

O'Reilly RC, Munakata Y, McClelland JL (2000) Computational explorations in cognitive neuroscience: understanding the mind. Cambridge: MIT Press

O'Reilly RC, Frank MJ (2006) Making working memory work: a computational model of learning in the prefrontal cortex and basal ganglia. Neural Comput 18:283–328

Parasarathy H, Schall J, Graybiel A (1992) Distributed but convergent ordering of corticostriatal projections: analysis of the frontal eye field and the supplementary eye field in the macaque monkey. J Neurosci 12:4468–4488

Pasupathy A, Miller EK (2005) Different time courses of learning-related activity in the prefrontal cortex and striatum. Nature 433:873–876

Rainer G, Rao SC, Miller EK (1999) Prospective coding for objects in the primate prefrontal cortex. J Neurosci 19:5493–5505

Reynolds JN, Hyland BI, Wickens JR (2001) A cellular mechanism of reward-related learning. Nature 413:67–70

Robbins TW (2005) Chemistry of the mind: neurochemical modulation of prefrontal cortical function. J Comp Neurol 493:140–146

Roberts WA, Mazmanian DS (1988) Concept learning at different levels of abstraction by pigeons, monkeys, and people. J Exp Psychol Anim Behav Proc 14:247–260

Sawamura H, Shima K, Tanji J (2002) Numerical representation for action in the parietal cortex of the monkey. Nature 415:918–922

Schoenbaum G, Setlow B (2001) Integrating orbitofrontal cortex into prefrontal theory: common processing themes across species and subdivisions. Learn Mem 8:134–147

Schultz W, Dickinson A (2000) Neuronal coding of prediction errors. Ann Rev Neurosci 23:473–500

Selemon LD, Goldman-Rakic (1985) Longitudinal topography and interdigitation of corticostriatal projections in the rhesus monkey. J Neurosci 5:776–794

Shelton C (2000) Morphable surface models. Intl J Computer Vis 38:75–91

Sigala N, Logothetis NK (2002) Visual categorization shapes feature selectivity in the primate temporal cortex. Nature 415:318–320

Tanaka K (1996) Inferotemporal cortex and object vision. Ann Rev Neurosci 19:109–139

Tomita H, Ohbayashi M, Nakahara K, Hasegawa I, Miyashita Y (1999) Top-down signal from prefrontal cortex in executive control of memory retrieval [see comments]. Nature 401:699–703

Tsao DY, Freiwald WA, Tootell RB, Livingstone MS (2006) A cortical region consisting entirely of face-selective cells. Science 311:670–674

Van Hoesen GW, Yeterian EH, Lavizzo-Mourey R (1981) Widespread corticostriate projections from temporal cortex of the rhesus monkey. J Comp Neurol 199:205–219

Vogels R (1999a) Categorization of complex visual images by rhesus monkeys. Part 1: behavioural study. Eur J Neurosci 11:1223–1238

Vogels R (1999b) Categorization of complex visual images by rhesus monkeys. Part 2: single-cell study. Eur J Neurosci 11:1239–1255

Wallis JD, Miller EK (2003) From rule to response: neuronal processes in the premotor and prefrontal cortex. J Neurophysiol 90:1790–1806

Wallis JD, Anderson KC, Miller EK (2000) Neuronal representation of abstract rules in the orbital and lateral prefrontal cortices (PFC). Soc Neurosci Abs 365.5: 976

Wallis JD, Anderson KC, Miller EK (2001) Single neurons in the prefrontal cortex encode abstract rules. Nature 411:953–956

White IM, Wise SP (1999) Rule-dependent neuronal activity in the prefrontal cortex. Exp Brain Res 126:315–335

Wilson C, Kawaguchi Y (1996) The origins of two-state spontaneous membrane potential fluctuations of neostriatal spiny neurons. J Neurosci 16:2397–2410

Wood JN, Knutson KM, Grafman J (2005) Psychological structure and neural correlates of event knowledge. Cereb Cortex 15:1155–1161

Wyttenbach RA, May ML, Hoy RR (1996) Categorical perception of sound frequency by crickets. Science 273:1542–1544

Xu F, Spelke ES (2000) Large number discrimination in 6-month-old infants. Cognition 74:B1-B11

Yeterian EH, Van Hoesen GW (1978) Cortico-striate projections in the rhesus monkey: the organization of certain cortico-caudate connections. Brain Res 139:43–63

Molecules that Disrupt Memory Circuits in Alzheimer's Disease: The Attack on Synapses by Aβ Oligomers (ADDLs)

William L. Klein[1], *Pascale N. Lacor*[1], *Fernanda G. De Felice*[1], and *Sergio T. Ferreira*[1]

Summary

Individuals with early Alzheimer's disease (AD) suffer from a selective and profound failure to form new memories. A novel molecular mechanism with implications for therapeutics and diagnostics is now emerging in which the specificity of AD for memory derives from disruption of plasticity at synapses targeted by neurologically active Aβ oligomers. We have named these oligomers "ADDLs" (for pathogenic Aβ-derived diffusible ligands). ADDLs constitute metastable alternatives to the disease-defining Aβ fibrils deposited in amyloid plaques. In AD brain, ADDLs accumulate primarily as Aβ 12-mers ($\sim$54 kDa). The same size oligomers occur in tg-mouse AD models; in mice, these 12-mers appear concomitantly with memory failure, consistent with the ability of ADDLs to inhibit long-term potentiation (LTP) and block reversal of long-term depression (LTD). Mechanistically, ADDLs are gain-of-function ligands that bind with specificity to particular synapses, targeting synaptic spines. Binding leads to a rapid and ectopic expression of the memory-linked immediate early gene Arc. Such aberrant accumulation has been linked by others to memory dysfunction in tg-Arc mouse models. Consistent with the expected consequences of Arc overexpression, ADDLs promote loss of surface NMDA receptors and anomalous spine morphology, which are responses expected to contribute to plasticity failure and memory dysfunction. Importantly, the attack on synapses provides a putative mechanism that unifies AD memory dysfunction with major features of AD neuropathology. Recent findings show ADDL binding instigates synapse loss, AD-type tau hyperphosphorylation, and generation of reactive oxygen species (ROS). Binding sites for ADDLs are at or in the close vicinity of NMDA receptors. Antibodies against external domains of NMDA receptors reduce ADDL binding and inhibit ADDL-stimulated ROS formation. The ROS response also is inhibited by memantine, an open-channel blocker of NMDA receptors recently approved for AD therapeutics. The ability of memantine to contravene the impact of ADDLs offers a new mechanism to explain why an NMDA receptor antagonist should improve memory function in AD patients. Elimination of ADDLs by vaccines now under development could provide the first AD treatments that are truly disease-modifying. In addition to establishing a molecular mechanism of significant value for AD therapeutics and diagnostics, studies of ADDL interactions with synaptic pathways and control mechanisms ultimately may provide new insights into the extraordinary complexities of physiological synaptic information storage.

---

[1] Cognitive Neurology & Alzheimer's Disease Center, Department of Neurobiology and Physiology Northwestern University Institute for Neuroscience, 2205 Tech Drive, Evanston, IL 60208, USA, wklein@northwestern.edu

Bontempi et al.
Memories: Molecules and Circuits
© Springer-Verlag Berlin Heidelberg 2007

Individuals with early Alzheimer's disease (AD) specifically suffer from a profound failure to form new memories. As the disease progresses, retrieval of stored memories also fails, and the dementia grows increasingly broad and catastrophic. A sense of what this means for people with the disease comes from a story involving Auguste D, the subject in the first case study of AD 100 years ago (Alzheimer 1906). Auguste D was asked in a visit by Dr. Alzheimer to write her name. Unable to do so, she said instead, "I have lost my self...". Fortunately, in this centennial year of Alzheimer's seminal study, the means to block this profound loss of self may not be far away. We can begin to envision disease-modifying therapeutics that might stop progression to end-stage disease, perhaps even restoring memory function to normalcy in early-stage disease. While these prospects appear optimistic, they stem from increasingly instructive insights into the molecular mechanisms of AD pathogenesis.

As we attempt to define disease mechanisms, there are two critical benchmarks that must be satisfied. Ultimately, the correct theory must explain why early AD targets memory and it must account for AD's pathological hallmarks. Satisfying these benchmarks is challenging, and at times it has been suggested that AD could be multiple diseases. Nonetheless, as reviewed below, evidence is accumulating to indicate that memory loss and neuropathology might be attributable to the toxic impact of soluble oligomers of Aβ, the same 42-amino acid peptide that generates the fibrils found in amyloid plaques. Much smaller than the large insoluble fibrils, the oligomers are just a little larger than 50 kDa in mass, i.e., similar to an average-sized globular protein. As reviewed below, Aβ oligomers attack synapses with great specificity, disrupting synapse shape, composition and the mechanisms of plasticity. Newer findings show that oligomers also trigger neuronal damage characteristic of the major features of AD neuropathology: synapse loss, tau hyperphosphorylation, oxidative stress, and ultimately nerve cell death. Oligomers thus seem to be appealing candidates for the hidden toxins of AD memory loss and neuropathology.

The original amyloid cascade hypothesis: fibrils and nerve cell death

Early findings from nerve cell biology, human genetics and neuropathology provided an exceptionally strong foundation for the concept that Aβ-derived fibrillar toxins play a primary role in AD pathogenesis. This hypothesis, summarized in 1992 in a seminal review by Hardy and Higgins, has been extraordinary in its ability to generate experimental progress. The benchmark evidence for the amyloid cascade (coupled with more than 15,000 citations to Aβ in PubMed) indeed seems compelling:

- AD is linked by human genetics to the Aβ peptide.
- Aβ in vitro self-assembles into lethal neurotoxins.
- toxic Aβ preparations contain abundant, readily-detectable fibrils.
- fibrils formed in vitro mimic those found in AD plaques.
- AD is accompanied by significant nerve cell death.

These findings generated a simple and persuasive concept – AD is, in essence, the consequence of neuron death induced by insoluble deposits of large amyloid fibrils. The original amyloid cascade has been a powerful hypothesis, but despite its compelling support, it misses the mark in explaining memory loss in early AD. Critical flaws are epitomized by findings in two transgenic mouse vaccine studies published in 2002. These studies (Dodart et al. 2002; Kotilinek et al. 2002) tested the therapeutic impact of monoclonal antibodies against Aβ. Consistent with the amyloid cascade hypothesis, the mouse models expressing the human amyloid precursor protein (APP) manifested both age-onset plaques and memory failure. However, the vaccination produced results that were strikingly inconsistent with the hypothesis. First of all, memory loss was found to be reversible. One study showed reversal after about two weeks (Kotilinek et al. 2002), the other in as little as 24 hours (Dodart et al. 2002). Secondly, despite reversing memory loss, the Aβ antibodies did not reduce amyloid plaque burden. These models of early AD thus manifest an Aβ-dependent memory loss that is not the consequence of neuron death and is not instigated by the presence of amyloid plaques. In fact, the lack of correlation between plaque burden and dementia in humans has long been noted by neuropathologists (Katzman et al. 1988), who have considered this a major flaw in the original amyloid cascade hypothesis.

The alternative to fibrillar toxins: neurological damage by soluble Aβ oligomers

If plaques do not initiate the disease, how does one explain the consequences of mutated APP or the therapeutic efficacy of anti-Aβ monoclonals? Newer findings indicate that the crucial molecules are not found in senile plaques and that the neurological impact of these molecules significantly precedes neurodegeneration. While retaining the central role for Aβ-derived toxins, these findings strongly support a new concept: early memory loss comes from failure of synaptic plasticity, not neuron death, and the fundamental molecular pathogens are small soluble Aβ oligomers, not fibrils. In other words, fibrils are not the only toxins derived from Aβ, and probably not the most important ones. How these concepts emerged and underlie new opportunities for early-stage diagnostics and disease-modifying therapeutics is reviewed in the remainder of this chapter.

An early clue came from experiments by Finch and colleagues (Oda et al. 1994, 1995) on the interaction of Aβ with ApoJ (clusterin), an upregulated, plaque-associated molecule in AD brain. Rather than promoting plaque formation, ApoJ proved surprisingly effective at blocking Aβ from forming large aggregates. Inhibition of aggregation occurred even at substoichiometric doses of ApoJ (1 part ApoJ, 20 parts Aβ), suggesting a chaperone-like action. Toxicology experiments with the PC12 pheochromocytoma cell line provided an even bigger surprise. ApoJ-Aβ solutions, which lacked sedimentable aggregates of Aβ, remained highly effective at blocking MTT reduction, an indicator of vital mitochondrial function and vesicle trafficking. The ApoJ experiments provided the first indication that Aβ-containing structures other than large insoluble fibrils might be pathogenic.

A different outcome for Aβ self-assembly: globular molecules rather than fibrillar aggregates

The nature of the Aβ assemblies produced in the presence of ApoJ was first investigated by our group (Lambert et al. 1998), leading to the discovery that toxin-producing Aβ self-assembly does not lead inexorably to large fibrillar aggregates or even protofibrils. An alternative outcome comprises small globular molecules only several nanometers in diameter. Atomic force microscopy (AFM) images of ApoJ-Aβ solutions ruled out the presence of fibrils or short rod-like protofibrils, showing only small structures roughly comparable in dimensions to soluble globular proteins smaller than 100 kDa. Unlike the readily observed fibrils, such nanoscale structures would be cryptic to conventional neuropathology and easily missed in electron microscopy.

Alternative methods without ApoJ have been developed to generate fibril-free, globular Aβ assemblies. In our original approach (Lambert et al. 2001; Klein 2002), Aβ42 was monomerized by HFIP, dissolved in fresh DMSO, diluted into cold F12 culture medium, and briefly centrifuged. The supernatants were completely free of protofibrils (Chromy et al. 2003). Even after 24 hours at 37 degrees, micromolar solutions exhibited no rod-like protofibril structures, establishing that the globular molecules seen in AFM were at least metastable. Most importantly, these ApoJ-free solutions were neurotoxic, proving that the active molecules were homo-oligomers of Aβ, not ApoJ- Aβ complexes.

Globular toxins are built of SDS-resistant oligomers

The composition of the globular neurotoxins has been analyzed by gel chromatography and electrophoresis (Lambert et al. 2001; Chromy et al. 2003). SDS-resistant Aβ42 oligomers typically range from dimer to 24-mer. Aβ40, which physiologically is much more abundant than Aβ42, does not generate stable oligomers (Levine 1995, 2004). The spectrum of oligomers resolved by electrophoresis is much more evident by Western blot than silver stain, although Western blot patterns are markedly antibody-dependent. The commonly used 4G8 monoclonal, for example, is unusually sensitive to dimers while poorly recognizing mid- to large-sized oligomers (Lambert et al. 2001; Chromy et al. 2003). Other experimental conditions also influence the analysis. For example, some oligomers are stable in SDS at room temperature, but they break down when subjected to boiling, a step often used in SDS-PAGE. Although it has been suggested that SDS-PAGE itself might generate small oligomers, in fact monomers sans oligomers are readily detected by SDS-PAGE using fresh preparations (Chromy et al. 2003). Oligomerization occurs quickly, however, even at low concentration. Oligomers form from 10 nM solutions of monomer within minutes (Chang et al. 2003), and the process of immunoblotting will generate oligomers from concentrated monomers if SDS is not constantly present.

Not all oligomeric states are equally abundant, and oligomer assembly itself is extremely sensitive to in vitro conditions, presumably reflecting the conformational dynamics of Aβ. Smaller SDS-resistant species (3- and 4-mers) are favored in solutions kept at 4 degrees, but stable 12-mers become prominent when dilute solutions of tetramers are incubated at 37 degrees (Klein 2002). Conformation differences, moreover, exist between oligomers of the same size. For example, SDS-stable oligomers have been observed that are not toxic and are unreactive with antibodies that bind toxic species (Chromy et al. 2003). Physiologically relevant factors also have been identified that affect oligomerization, including the oligomer-promoting action of divalent cations (Huang et al. 2004) and levuglandin (Boutaud et al. 2005). Overall, however, the mechanism of oligomer formation, the stability of oligomers, and the relationship between oligomerization and fibrillogenesis still remain poorly characterized.

Neurological impact: oligomers rapidly block synaptic information storage

Metastable Aβ ogligomers have generated major interest because of their remarkable impact on mechanisms germane to memory loss. Memory formation begins at synapses, so one might anticipate that destruction of memory formation in AD itself begins at synapses. A crucial question, therefore, has been whether oligomers might compromise synaptic plasticity. The answer has been clear-cut. In LTP and LTD, the classic paradigms for learning and memory, Aβ oligomers are neurologically potent CNS toxins that rapidly disrupt synaptic information storage.

Initial experiments (Lambert et al. 1998; Klein 2001) tested stereotaxic injections of ApoJ-derived oligomers on LTP in living mice. Although neither ApoJ alone nor amyloid fibrils had any impact, the injected oligomers produced a rapid and profound inhibition. Inhibition was confirmed and extended in brain slice experiments in which oligomers prepared without ApoJ completely block LTP at sub-micromolar concentrations (Lambert et al. 1998). Effects occurred within an hour and were selective for particular aspects of plasticity. While oligomers inhibit LTP, they do not inhibit LTD (Wang et al. 2002). On the other hand, oligomers do block the reversal of LTD. The net neurological effect of oligomers in the hippocampus thus is to repress positive synaptic feedback. To account for these altered states of plasticity, it has been hypothesized that oligomers disrupt glutamate receptor trafficking (Gong et al. 2003; Lacor et al. 2004a), which is essential for both LTP maintenance and LTD reversal (Sheng and Lee 2001). Recent evidence supports this possibility (Lacor et al. 2004b, 2005; Snyder et al. 2005). A mechanism to account for disrupted trafficking is discussed at the end of the article.

The initial impact of oligomers on synaptic plasticity is rapid and selective. There is no effect, for example, on evoked action potentials, so the mechanism is non-degenerative, at least at first. These findings led to a prediction that was somewhat iconoclastic: if oligomers were responsible for memory loss in early AD through an impact on synaptic plasticity, then early memory loss should be reversible (Lambert et al. 1998; Klein 2001). As discussed earlier, this prediction was verified in the impressive vaccination experiments on mouse models (Dodart et al. 2002; Kotilinek et al. 2002).

The amyloid cascade hypothesis revisited: memory loss in early AD is a synaptic disease caused by soluble Aβ oligomers

LTP is not memory but is widely recognized as a good experimental paradigm for the study of memory mechanisms. The rapid impact of oligomers on LTP thus is intuitively appealing in its relevance to AD, and in 1998 we proposed a new hypothesis that attributed early memory loss to oligomer-induced failure in synaptic plasticity (Lambert et al. 1998). The experimental foundation for this concept has been substantiated in multiple investigations (Chen et al. 2000; Vitolo et al. 2002; Klyubin et al. 2004, 2005; Wang et al. 2004a; Costello et al. 2005; Nomura et al. 2005; Puzzo et al. 2005; Trommer et al. 2005; Walsh et al. 2005), with especially strong support found in a major study of LTP by Walsh, Selkoe and colleagues (Walsh et al. 2002). That group showed that oligomers found in medium conditioned by hAPP-transfected CHO cells are exceptionally potent inhibitors of LTP in vivo. Western blots indicate that the cell-derived oligomers comprise mostly small SDS-stable oligomers, free of structures that might be considered protofibrils, although given antibody selectivity and gel conditions it is unclear whether mid-sized oligomers might also be present (e.g., 9-24-mers). As predicted, controls using insulin-degrading enzyme, which proteolyzes monomers but not oligomers, ruled out the possibility that monomeric Aβ contributed to inhibition, whereas drugs that blocked Aβ production also blocked accumulation of the neurologically active molecules.

The compelling verification that oligomers are neurologically significant and sufficiently stable to accumulate in a cell-conditioned media, along with the vaccine results validating the prediction that memory loss is reversible, provided impetus for significant modifications in the amyloid cascade hypothesis (Hardy and Selkoe 2002). The new cascade includes two significant emendations: 1) early memory loss now is attributed to synapse failure, not neuron death, and 2) synapse failure is attributed to Aβ oligomers, not amyloid fibrils.

The hypothesis that Aβ oligomers play a role in AD pathogenesis has stimulated interest in the broader possibility that toxic protein oligomers may be common to multiple diseases (Klein 2005). Aβ is one of more than 20 different fibrillogenic proteins that are disease-linked and, until recently, it had been assumed that the pathogenic molecules were fibrillar. However, a number of these proteins now have been found to generate sub-fibrillar, oligomeric cytotoxins (Conway et al. 2000; Butler et al. 2003; Reixach et al. 2004). In some cases, contrary to earlier dogma, oligomers exhibit cytotoxicity but fibrils do not. Extrapolation from Aβ as a specific case study thus is providing general insight into diseases of protein folding and mis-assembly. Strategies that target elimination of soluble oligomeric toxins could provide new therapeutics for a broad range of significant diseases.

Clinical substantiation that AD pathology includes neurologically active Aβ oligomers

If soluble Aβ oligomers are to be valid targets for AD therapeutics, they must be present in human brain and manifest a strong AD-dependent accumulation. All the evidence reviewed so far comes from experimental models and oligomeric toxins generated in vitro. The critical question is whether AD-affected brains contain identical neurologically active Aβ oligomers.

Early evidence indicated that pre-fibrillar assemblies indeed existed in AD-affected brains (Frackowiak et al. 1994; Kuo et al. 1996). Since fibrils were regarded as the initiators of pathogenesis, it was assumed the oligomers simply were surrogates for ongoing fibrillogenesis. The oligomers were considered transient intermediates en route to formation of the pathogenically relevant amyloid fibrils. However, given the potential of the oligomer hypothesis to account for AD memory loss, significant efforts recently have sought to substantiate the AD-dependent accumulation of brain oligomers.

Because not all oligomers are toxic (Chromy et al. 2003), it has been important to determine whether brain-derived oligomers are structurally equivalent to the neurologically disruptive, laboratory-derived oligomers. The key tools for rigorous analysis have been conformation-dependent antibodies. Such antibodies are readily generated by oligomer-based vaccines (Lambert et al. 2001), which are highly immunogenic and superior to short peptides for generating antibodies recognizing conformational epitopes. Polyclonal antibodies have been obtained that are at least 1000 times more sensitive for oligomers than monomers (Lambert et al. 2001; Chang et al. 2003) and are capable of detecting less than 0.1 fmole oligomerized Aβ in dot immunoblots. Monoclonal antibodies have been obtained with similar properties (Lambert et al. 2003, Lambert et al. 2007), whereas some show unique attributes. One interesting monoclonal, for example, binds oligomers but not fibrils (Chromy et al. 2003), whereas another binds fibrils but not oligomers (Lambert et al. 2007), suggesting that fibrils are not simple assemblies of oligomeric subunits. A second approach to antibody generation by Glabe and colleagues has been to use oligomers coupled to gold particles (Kayed et al. 2003). These immunogens have generated oligomer-selective antibodies with an intriguing ability to recognize oligomers of multiple fibrillogenic proteins besides Aβ (e.g., alpha synuclein and IAPP); it has been proposed, therefore, that particular structural domains may be common to toxic protein oligomers in general (Kayed et al. 2003). The generic oligomer antibodies do not bind fibrils, consistent with the concept that structures of non-fibrillar and fibrillar toxins are not related in a simple manner.

Oligomers occur in human brain and are strikingly elevated in AD

Assays based on oligomer-selective antibodies have provided compelling evidence that oligomers are bona fide components of AD pathology: oligomers occur in human brain and are strikingly elevated in AD; oligomers from AD brain are structurally homologous to synthetic oligomers; and hAPP mouse models of AD accumulate oligomers as predicted. Oligomers found in CSF, moreover, appear to provide a significant biomarker of AD that is potentially of value for future clinical diagnostics.

To assess soluble oligomers, brain extracts have been prepared using physiological buffers to avoid chemical perturbation. Soluble fractions, obtained after ultracentrifugation and analyzed by dot immunoblots, show a remarkable, AD-dependent increase in oligomers. AD subjects show levels up to 70-fold higher than those found in controls (Gong et al. 2003), verified with three different oligomer-selective antibodies (Kayed et al. 2003; Lacor et al. 2004a). The large increase in extractable oligomers is regionally selective, robust in frontal cortex but marginally detectable in cerebellum (Lacor et al. 2004a; Fig. 1), a pattern consistent with other aspects of AD pathology.

With respect to structure, three types of data indicate that soluble oligomers in brain extracts are equivalent to the toxic oligomers prepared in vitro. First, brain oligomers react with conformation-dependent antibodies raised against synthetic oligomers. These conformation-dependent antibodies not only distinguish oligomers from monomers but also distinguish toxic oligomers from non-toxic species; neither capacity is typical of antibodies generated against short peptides (Lambert et al. 2001). Second, synthetic and AD brain-derived preparations show equivalence by two-dimensional gel analysis. Each contains a prominent oligomer with isoelectric point of $\sim$ 5.6 and SDS-stable mass of $\sim$ 54 kDa, i.e., 12-mers (Gong et al. 2003). Additional species have been obtained by extracting brain with SDS, indicating some oligomers are trapped or tightly bound. Third, in cell culture experiments discussed below, both synthetic and soluble AD brain-derived oligomers show attachment to nerve cell surfaces in identical, ligand-like patterns (Gong et al. 2003; Lacor et al. 2004a). No binding occurs with extracts from control brain. Overall, the data indicate that neurologically active oligomers formed in vitro are structural homologs of oligomers that accumulate in AD brain.

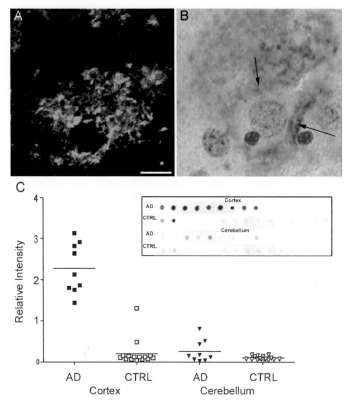

Fig. 1. Aβ oligomers are deposited extracellularly around neurons and are highly elevated in Alzheimer's-affected cortex. Clinical substantiation of ADDL involvement in AD pathology has come from immunomicroscopy and immunochemistry. Cortical sections from AD brain treated with ADDL-generated antibodies show diffuse synaptic-type labeling surrounding individual pyramidal neurons (fluorescent- or peroxidase-conjugated secondary antibodies, (**A**) and (**B**), respectively). Similar staining is evident in MCI (mild cognitive impairment). Immunodot-blots of soluble brain extracts confirm the AD-dependent accumulation of ADDLs (**C**). ADDL pathology is pronounced in cortex but not cerebellum. (Reprinted from Lacor et al. 2004a)

TG mouse models of AD accumulate soluble oligomers

Animal models of AD are essential for studies of mechanism and drug discovery, so it has been important to determine whether they also accumulate oligomers. Early evidence suggested indirectly that oligomers should be present. For example, some hAPP transgenic mouse strains manifest brain damage (synapse loss, deficiencies in LTP, poor memory performance) that does not correlate with plaque burden, in some cases despite the absence of amyloid deposits. These animals essentially recapitulate the poor correlation between plaques and dementia in humans, but in an extreme manner. Although originally rejected as inadequate models because they lacked amyloid plaques, some of these strains in fact may be excellent models for early AD caused by toxic oligomers.

In harmony with human data, the occurrence of oligomers has recently been confirmed in multiple strains of tg mice. Even when the commonly used 4G8 antibody shows no increase, assays with oligomer-selective antibodies detect increases in immunoreactivity that reach several-hundred-fold over controls (Chang et al. 2003). These robust increases in oligomers are regionally selective, indicating that broad diffusion does not occur. In the highly regarded triple transgene model of Oddo et al. (2003), which exhibits AD-like tau phosphorylation as well as plaques, two different oligomer-selective antibodies showed the same pattern of developmental increase, with a transient peak in oligomers at six months and a sustained elevation after 15 months (Oddo et al. 2006).

Quantitative correlation of oligomer levels with memory dysfunction was first evident in experiments done in collaboration with Karen Ashe's group (Westerman et al. 2002), in which memory loss occurred after oligomer levels increased past certain threshold levels, a possibility consistent with the concept of cognitive reserve (Mesulam 1999). More recently, that group has reported memory loss correlates best with the accumulation of SDS-stable 12-mers (Lesne et al. 2006). This important finding confirms that memory dysfunction in tg mouse models is linked to the same oligomers found by Gong et al. in AD brain. Another mouse model, developed by Ohno et al. (2005), also shows highly upregulated soluble oligomers; in these animals, knockout of β-secretase eliminates oligomers and blocks deficiencies in a fear-linked memory task.

The abundance of oligomers resulting from transgene expression in these animals likely explains why brain deficiencies correlate poorly with amyloid plaques, and it substantiates their value as models for early AD pathology and memory loss.

Human neuropathology: distribution of oligomers in situ is consistent with links to AD onset

Immunohistochemistry with oligomer-selective antibodies readily distinguishes AD brain sections from controls (Lambert et al. 2007). Besides confirming the presence of oligomers, results reveal an unexpected specificity to oligomer distribution in situ. In lightly stained AD brain sections, oligomers show a perineuronal localization (Lacor et al. 2004a), indicating the predominant localization is not cytoplasmic (Fig. 1). Although oligomers accumulate in the cytoplasm of certain tg mouse strains (Takahashi et al. 2004), this could be the consequence of very high oligomer production in mice. Oligomers also do not co-localize with the fibrils of compact plaques (Kayed et al. 2003). Perineuronal distribution likely corresponds to so-called diffuse deposits, a nomenclature that may be misleading with respect to cellular mechanisms. "Diffuse deposit" implies a non-selective extracellular precipitate, but the distribution surrounding individual neurons suggests that oligomers may associate with particular sites in dendritic arbors. Perineuronal distribution is readily observed for isolated, individual neurons, suggesting that the source of oligomers may be that same neuron.

Perineuronal/dendritic staining around isolated neurons appears to be the first sign of pathology in preclinical brain sections (Bigio et al. 2005). This fact, along with observations that soluble oligomers in control brain extracts sometimes show minor elevations (Gong et al. 2003), indicates that oligomer accumulation begins preclinically. The relationship between accumulation and dementia is unlikely to be straightforward, with current results indicating that dementia would manifest only after oligomer levels exceed some critical threshold. Recently developed plaque-imaging agents (Klunk et al. 2004) probably will be inadequate to detect oligomers at threshold levels because oligomer distribution is distinctly segregated from thioflavin-positive plaques (Kayed et al. 2003). However, alternative imaging agents that specifically target oligomers are under development (Montalto et al. 2004) and might provide a means for early AD detection.

Nanotechnology-based assays of cerebrospinal fluid confirm the extracellular presence of oligomers

The occurrence of oligomers in soluble brain extracts, along with their perineuronal distribution, provides indirect evidence that oligomers occur extracellularly and thus are potentially accessible for clinical assays. If extracellular localization exists, then oligomers should occur in CSF, but neither ELISAs nor dot immuno-blots verify this prediction, even using the most sensitive oligomer-generated antibodies. However, in a recent and unique approach using nanotechnology, we have succeeded in increasing sensitivity by orders of magnitude while still retaining immunospecificity (Georganopoulou et al. 2005). With this new assay, CSF oligomer levels show marked AD dependence, with median levels for AD subjects elevated 10-fold over controls (p < 0.0001 for 30 samples). Overlap between populations is virtually non-existent. Although a much larger sample is needed to validate these results, and the current assay itself is still being enhanced, the use of oligomers as biomarkers in AD diagnostics shows unique promise.

Cellular mechanism – oligomers specifically target synapses and disrupt the molecular cell biology of memory

Oligomers block synaptic information storage in electrophysiological experiments. They trigger AD-type pathology in transgenic models and likely instigate memory loss. Furthermore, in human brain, oligomers show a striking accumulation around neurons that begins in very early stages of AD. There clearly is a need to understand how oligomers act in terms of molecular mechanisms. How is it that oligomers attack neurons, and what memory-relevant molecular changes are triggered? An appealing hypothesis is suggested by the putative association of oligomers with dendritic arbors in situ: perhaps oligomers attack and disrupt signaling pathways specifically at sites critical to memory formation, in essence acting as pathogenic ligands.

Patterns in culture recapitulate patterns in vivo: oligomers are "Aβ-derived diffusible ligands" (ADDLs) that bind to hot-spots on dendritic surfaces

One would predict, if the ligand hypothesis were true, that incubation of cultured neurons with soluble oligomers from AD brains should generate a perineuronal pattern analogous to that observed in brain sections. Experiments with highly differentiated monolayers of rat hippocampal neurons, using oligomer-selective antibodies to localize distribution, have confirmed this prediction (Gong et al. 2003; Lacor et al. 2004a). Oligomers in crude extracts from AD brains show remarkably patterned attachment to neuron surfaces, localizing at small hot-spots that are especially rich in mature dendritic arbors. CSF from AD subjects also gives this pattern, although binding is faint, consistent with very dilute oligomer levels. Patterns of identical perineuronal hot-spots are generated when neurons are incubated with synthetic oligomers. Extracts from control brain, however, show no binding. The selective binding of Aβ oligomers, whether from AD brain or prepared in vitro, substantiates their ligand nature. This property, coupled with their neurological impact and their solubility, has led to the acronym "ADDL" (pronounced "addle"), for a pathogenic "Aβ-derived diffusible ligand".

The pattern of binding to cultured neurons supports the hypothesis that diffuse deposits in AD originate with oligomers bound to particular sites in dendritic arbors. If oligomers are gain-of-function pathogenic ligands, a key question is how specific is their target? In fact, in any given hippocampal culture, oligomers attack only a subset of neurons, typically between 25% and 50% of those present. Neurons side-by-side in culture can be ADDL-positive and ADDL-negative (Lacor et al. 2004a), even with similar morphology and expression of memory-relevant molecules such as CaMKII. What discriminates them is not yet known. Binding sites, however, are sensitive to low-doses of trypsin (Lambert et al. 1998), indicating the existence of particular proteins that serve adventitiously as oligomer receptors.

A key to memory loss - oligomers (ADDLs) specifically target synapses

The identity of the hot-spots themselves may be the key to memory loss mechanisms. More than 90% of ADDL hot-spots co-localize with puncta of PSD-95 (Lacor et al. 2004a), a major component of post-synaptic densities that, in mature hippocampal cultures, is a reliable marker for post-synaptic terminals (Allison et al. 2000). Colocalization of ADDLs with PSD-95 establishes that ADDLs are ligands that specifically target synapses. Roughly 50% of the synapses do not bind ADDLs, so specificity extends also to the nature of the synapses. The fact that ADDLs are ligands that directly and specifically attack synapses provides substantial strength to the hypothesis that memory loss could be an oligomer-induced failure of synaptic plasticity.

Importantly, synaptic binding is evident for ADDLs obtained from AD brain as well as prepared in vitro. Simple fractionation of brain extracts or synthetic ADDLs indicates the ligands are the 12-mers identified by two-dimensional gel analysis (Lacor et al. 2004a).

Sub-synaptic targeting: ADDLs bind to synaptic spines and disrupt translation of Arc, an immediate early gene essential for long-term memory formation

Confocal immunofluorescence microscopy shows that the synaptic targeting of ADDLs frequently localizes to dendritic spines, as illustrated here by a neuron double-labeled for ADDLs and CaMKII (Fig. 2). While other sites also may be targeted, spines are particularly interesting because they contain the post-synaptic signal transduction components of excitatory synapses. Signaling events within spines orchestrate the

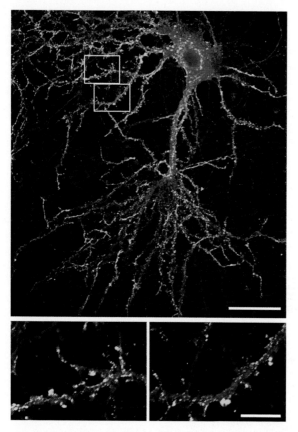

Fig. 2. Localization of ADDL binding sites to dendritic spines. ADDLs act as pathogenic gain-of-function ligands that target synapses. Here, ADDLs have been added to cultures of highly differentiated hippocampal cells (21 DIV), which then were immunolabeled for ADDLs (*green*) and αCaMKII (*red*). In overlaid images, ADDLs show a highly patterned distribution seen to coincide (*yellow*) with dendritic spines. More than 90% of the ADDL puncta occur at synapses. Scale bars: *Top*, 40 μm; *Bottom insets*, 8 μm. Adapted from (Lacor et al. 2004a)

molecular plasticity essential for initiating memory formation. CaMKII, for example, plays a critical role in synaptic information storage and is enriched in synaptic spines. An important feature of spines is their dynamic cytoskeleton (Carlisle and Kennedy 2005). Spines are rich in actin and actin-associated proteins, and this molecular machinery regulates spine geometry and also the trafficking of glutamate receptors, essentially the underpinnings of synaptic plasticity.

Experiments focusing on the relationship between ADDLs and an actin-associated protein in synaptic spines have proven especially salient. This protein is Arc (acronym for activity-regulated cytoskelet on protein), the product of an immediate early gene required for long-term memory formation (Guzowski et al. 2000)). Arc mRNA traffics specifically to spines, where its translation is induced by synaptic transmission in a transient manner (Steward and Worley 2002). Chronic over-expression has been predicted to impair long-term memory, essentially by increasing noise in information processing (Guzowski 2002) and, in Arc tg mice, chronically high levels of Arc in these animals correlate with poor performance in learning tasks (Kelly and Deadwyler 2003).

When neurons are exposed to ADDLs, Arc is ectopically induced. After five minutes, there is an induction of small Arc puncta that coincides with ADDL-labeled spines (Lacor et al. 2004a). This response, however, is not transient. After six hours, Arc immunoreactivity remains markedly elevated and spread ectopically throughout the dendritic arbor. ADDLs thus provoke chronic Arc overexpression, a specific molecular anomaly associated with dysfunctional learning in tg mice.

We have considered how dysfunctional learning might be caused by particular synaptic changes resulting from Arc overexpression. Since Arc is an F-actin associated protein, we predicted that its chronic elevation might cause aberrations in glutamate receptor trafficking and spine shape (Lacor et al. 2004a), factors linked to F-actin that are expected to be critically associated with synaptic plasticity. We previously suggested that receptor trafficking dysfunction might be related to the deficiencies in LTP maintenance and LTD reversal caused by ADDLs (Gong et al. 2003), and Kelly and Deadwyler (2003) have suggested the learning dysfunction found in their transgenic Arc mice is due to actin-dependent loss of spine structural plasticity.

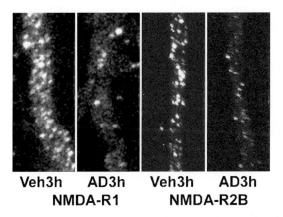

Veh3h   AD3h   Veh3h   AD3h
NMDA-R1   NMDA-R2B

Fig. 3. NMDA-R subunits NR1 and NR2B levels are reduced after ADDL treatment in rat primary hippocampal neurons. Twenty-one-day-old hippocampal neurons were exposed to 500 nM ADDL for three hr and fixed while permeabilization steps were excluded to ensure labeling of surface receptors. Labeling was done using antibodies raised against the extracellular domain of the receptor subunits NR1 and NR2B (ImmuQuest, 1:200 and Santa Cruz, 1:200, respectively). High magnification images of dendritic branches were collected by confocal microscopy, and the numbers of NMDA-R-immunoreactive puncta were quantified using Metamorph software (Meta Image Series, Universal Imaging Corporation). As shown in the representative image, the density of NR1- and NR2B-positive puncta (number of puncta per length of dendrites) was significantly reduced, by 70%, after ADDL treatment (p < 0.01)

Fig. 4. ADDL exposure affects drebrin-labeled dendritic spine morphology and density. Primary rat hippocampal neurons (21 days old) were treated with 500 nM ADDLs or with vehicle, fixed, and double-labeled as previously published (Lacor et al. 2004) for oligomer (M94, 1:2000) and for drebrin (Stressgen, 1:500). Representative images collected using a Leica LCS confocal microscope are reconstructions of a stack of z sections captured at 0.5-um intervals of a hippocampal neuron. Dendritic spine distribution (dendritic spines are identified as drebrin-positive puncta) after 5 min, 6 hr and 24 hr of ADDL (**A**, **B** and **C**) or after 24 hr of vehicle treatment (**D**). Spine deficits were detected in 6-hr and 24-hr ADDL-treated neurons compared to 5-min ADDL-treated and 24-hr vehicle-treated neurons. The right column represents magnified dendritic spines depicting mushroom-type or stubby spines (in controls) and dendritic protrusions that are markedly longer, reminiscent of dendritic filopodia, and sometimes branched protrusions. After 24 hr, severe spine disappearance is observed in ADDL-treated cells

Predictions validated: ADDLs alter spine composition and shape

We recently have validated the prediction that ADDLs exert a significant impact on both receptor levels and spine shape. Exposing neurons to ADDLs causes a 70% decrease in NR1 and NR2B subunits of NMDA-type glutamate receptors from the plasma membrane (Fig. 3; Lacor et al. 2004b, 2005, 2007). These receptors are critical for synaptic plasticity, and their down-regulation likely accounts for glutamate-initiated CREB phosphorylation following ADDL exposure (Tong et al. 2001). Interestingly, ADDLs also trigger a decrease in EphB receptors, which in spines are associated with NMDA receptor signaling. With respect to spine shape, ADDLs cause the short, club-like structure of controls to be supplanted by long, spindly spines (Fig. 4). This type of aberrant spine geometry is known to exhibit low glutamate receptor levels (Matsuzaki et al. 2001), consistent with the impact of ADDLs. Most significantly, spines that are aberrantly long and spindly have been linked to mental retardation and prionoses (Fiala et al. 2002). Although predicted from the Arc response, the mechanism underlying the ADDL-induced shift in spine shape is not yet known. EphB could be involved because of its established relation to spine morphogenesis and association with Rho/Rac signaling pathways, and, in fact, ADDLs affect Rac activity (Chromy et al. 2003). Additionally, the spine shape change induced by ADDLs is accompanied by rearrangements in drebrin and spinophilin, which are actin-regulating proteins. While Arc induction and its predicted consequences in response to ADDLs are very intriguing, oligomers also impact other important signaling pathways germane to synaptic plasticity (Tong et al. 2001; Vitolo et al. 2002; Wang et al. 2004a; Puzzo et al. 2005). Overall, a comprehensive understanding of the relationships between ADDLs and signaling remains an important goal.

Coupling of ADDLs to major features of AD neuropathology

Synapse loss and neuron death

Studies with transgenic animals have strongly suggested that ADDLs might be responsible for synapse elimination (Mucke et al. 2000), an aspect of neuropathology that in humans affords the best correlate of dementia (Masliah et al. 1993). This possibility has

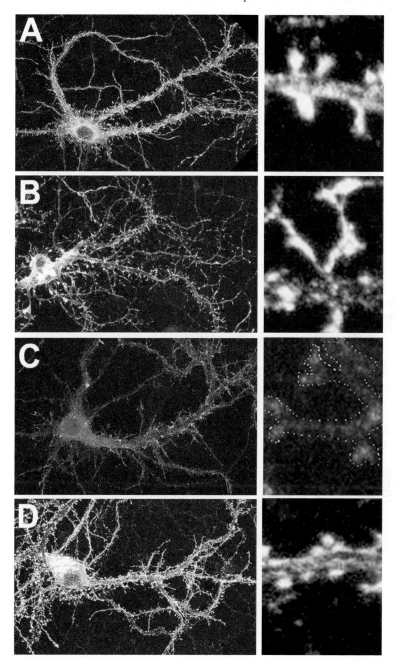

been verified directly in cell biology experiments. After 24 hours of exposure to AD-DLs, the spiny post-synaptic specializations manifest a significant decrease in density, with an overall loss of spines equaling 50%. Spine loss of this magnitude also has been reported for hAPP transgenic mice (Jacobsen et al. 2006). The original observations of synapse loss in transgenic mice correlated the pathology with soluble rather than in-soluble forms of Aβ (Mucke et al. 2000) and implicated a pathogenic role for the protein tyrosine kinase Fyn (Chin et al. 2005). Other tg mouse studies have reported decreased levels of Arc (Palop et al. 2005), consistent with degeneration of synaptic terminals. Interestingly, an early stage of synapse degeneration in AD may include adaptive en-largement of pre-synaptic terminals (Mukaetova-Ladinska et al. 2000; Scheff and Price 2003), a response also observed in hippocampal neurons exposed to ADDLs (Fig. 5). After short-term exposure, neurons exhibit bigger boutons as well as longer spines. At longer times, when spine levels have decreased 50%, the targeted neurons show a near-complete elimination of drebrin, another F-actin binding protein typically localized to spines (Fig. 4). Abnormalities in drebrin similarly manifest in hAPP transgenic mice (Calon et al. 2004).

The degeneration of synapses ultimately is superseded by neuronal degeneration and death, observed in organotypic slice cultures (Lambert et al. 1998). Neuronal loss is a significant characteristic of AD. In culture, knockout of Fyn is neuroprotective, indicating the pathogenic role of Fyn is shared in mechanisms responsible for death as well as synapse loss.

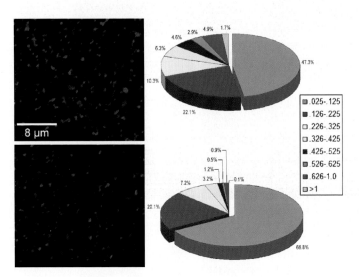

Fig. 5. ADDLs induce a shift toward bigger synaptic boutons. Images obtained by confocal microscopy of 6-hr ADDL (500 nM) and vehicle-treated hippocampal neurons illustrate that ADDLs increase the size of synaptophysin (Sigma, 1:500) immunolabeled puncta throughout the neuropil. Total number of pixels/field captured increased significantly after ADDL treat-ment (p < 0.02); however, this increase was not accompanied by an increase in the number of puncta/field (p > 0.05). The increase in the total number of labeled pixels/field can be explained by a significant shift toward larger sized puncta (p < 0.002). The percentage distributions of SVP-positive puncta in the different size categories are represented in the three-dimensional pie charts obtained from 10 different fields collected per group of treatments. Synaptically bound ADDLs might be involved in reactive synaptogenesis

Tau phosphorylation

In the revised amyloid cascade hypothesis, Aβ oligomers ultimately instigate formation of neurofibrillary tangles (Hardy and Selkoe 2002). Increased brain levels of soluble Aβ directly correlate with neurofibrillary tangle (NFT) density in AD patients (McLean et al. 1999) and Aβ oligomers have been shown to activate glycogen synthase kinase-3β (Hoshi et al. 2003), one of the kinases that appears to be involved in pathological tau hyperphosphorylation. However, direct demonstration that tau hyperphosphorylation could indeed be induced by Aβ oligomers has been lacking. Our recent results show that both synthetic and AD brain-derived ADDLs instigate marked increases in neuronal levels of AD-type tau phosphorylation, establishing that ADDLs are also coupled to this major facet of AD pathology (Fig. 6).

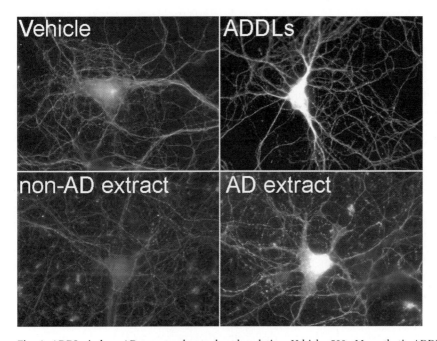

Fig. 6. ADDLs induce AD-type tau hyperphosphorylation. Vehicle, 500 nM synthetic ADDLs, 1 mg/ml soluble extract from non-AD or 1 mg/ml soluble extract from AD brains were added for four hours at 37 °C to cultures of highly differentiated rat hippocampal cells (21 DIV) grown in neurobasal medium supplemented with B27. After fixation, cells were immunolabeled overnight at 4 °C with phospho-Thr231 antibody (1:500) diluted in blocking solution. Neurons were then rinsed three times with PBS and incubated with Alexa Fluor 488 anti-rabbit IgG secondary antibody (1:1,000) diluted in blocking solution for three hours at room temperature. Cells were rinsed five times with PBS, coverslips were mounted with Prolong (Molecular Probes), and cells were visualized on a Nikon Eclipse TE 2000-U fluorescence microscope. Images were digitally acquired using MetaMorph software (Universal Imaging Corporation). Quantitative analysis of the data from multiple experiments, using Image J software (NIH Windows version), indicated that both synthetic ADDLs and AD brain-derived ADDLs increased tau hyperphosphorylation at Thr231 in hippocampal neurons by approximately three-fold

Blocking tau hyperphosphorylation has been achieved with a monoclonal anti-body that targets Aβ oligomers (but not monomers), which also blocks ADDL attachment to clustered synaptic binding sites. Abnormal tau hyperphosphorylation also was blocked by the Src family tyrosine kinase inhibitor, 4-amino-5-(4-chlorophenyl)-7(t-butyl) pyrazol (3,4-D) pyramide (PP2), and by the phosphatidylinositol 3-kinase inhibitor, LY 294002. Fyn kinase, a member of the Src family, has been linked to tau phosphorylation stimulated by insulin (Lesort et al. 1999) and by toxic preparations of Aβ (Zhang et al. 1996). Fyn signaling also reportedly underlies synaptic and cognitive impairments in a transgenic model of AD (Chin et al. 2005). Since ADDL-induced tau hyperphosphorylation in hippocampal cultures was manifested in the absence of neuronal death, these ADDL-stimulated signaling events may be an early event in AD neuronal dysfunction and degeneration.

Oxidative stress and the cellular basis of memantine action

We recently identified a fourth important facet of AD-relevant pathology induced by ADDLs, namely that ADDL attachment to synapses in hippocampal cultures stimulates formation of reactive oxygen species (ROS; Fig. 7). Oxidative damage is a significant attribute of the AD-affected brain. Because oxidative stress may result from the over-activation of NMDA-type glutamate receptors, we tested whether NMDA receptors were involved in ADDL-stimulated ROS formation. As shown (Fig. 7), ROS formation was completely blocked by memantine, an open-channel blocker of NMDA receptors. Inhibition of ADDL-induced ROS generation also was observed with another NMDA channel blocker, MK-801 (Fig. 7), and with antibodies against the extracellular domain of NR1 NMDA receptor subunits.

Since ROS at low physiological levels are necessary components of LTP signaling, it is plausible that anomalous over-stimulation could impair mechanisms relevant to memory formation. There is evidence, moreover, that high ROS levels are damaging in age-related impairment of LTP (reviewed in Serrano and Klann 2004). Therefore, the fact that ADDLs instigate marked elevations in neuronal ROS may provide an additional mechanism by which ADDLs deregulate signaling pathways involved in memory formation. This mechanism may be relevant to the therapeutic action of memantine, which recently was approved for AD patients. Because NMDA receptor function itself is crucial for memory formation, it has been a mystery why a receptor antagonist should be beneficial. By showing that memantine contravenes the pathogenic action of ADDLs, we have established for the first time a pathologically specific mechanism for the therapeutic value of memantine (Namenda).

ADDLs at synapses – a hypothetical basis for early AD, its diagnosis and treatment

We began with the premise that a unifying theory for AD must explain why its earliest stages selectively target memory formation and must also account for the major facets of AD neuropathology. Findings reviewed here strongly support the hypothesis that specific loss of memory function in early AD is the consequence of targeting and functional disruption of particular synapses by ADDLs. In the ADDL hypothesis (Fig. 8),

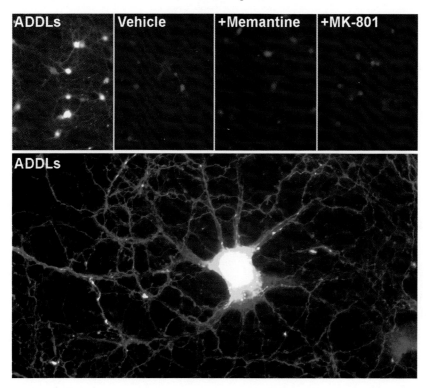

Fig. 7. ADDLs induce neuronal reactive oxygen species (ROS) generation. Vehicle, 500 nM AD-DLs, 500 nM ADDLs + 10 μM memantine or 500 nM ADDLs + 5 μM MK801 were added for three hours at 37 °C to cultures of highly differentiated rat hippocampal cells (21 DIV, grown in neurobasal medium supplemented with B27). Generation of ROS was monitored using the ROS-sensitive fluorescence dye, $CM-H_2DCFDA$ (5 μM, 40 minutes of probe loading). Neurons were rinsed three times with warm PBS, then two times with neurobasal media without phenol red. Cells were placed in neurobasal media without phenol red and then immediately visualized on a Nikon Eclipse TE 2000-U fluorescence microscope. Images were digitally acquired using MetaMorph software (Meta Image Series, Universal Imaging Corporation). Quantitative analysis of DCF fluorescence data obtained from multiple experiments, using Image J software (NIH Windows version), established that ADDLs increased ROS generation in hippocampal neurons by 3.5-fold and that this increase was fully blocked by memantine and MK801

oligomerization of Aβ42 provides gain-of-function pathogenic ligands that bind specifically to surface membrane proteins localized in synaptic spines. The affected synapses are those critically placed in memory circuits, and they are targeted because of localized ADDL production as well as adventitious ADDL capture by synapticallylocalized toxin receptors. As pathogenic agonists, ADDLs disrupt proper regulation of the spine cytoskeleton and thereby disrupt spine geometry and receptor trafficking. These spine deficiencies impair synaptic information storage. When the net impact of deficiencies in spine cell biology passes a critical threshold, the result is failed memory formation. Moreover, results that strongly substantiate the ADDL hypothesis have been obtained,

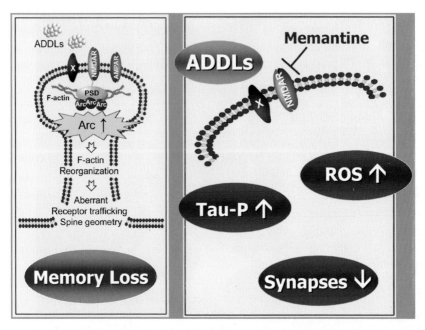

Fig. 8. Attack on synapses by ADDLs – a unifying mechanism to account for the specificity of AD for memory loss and major facets of neuropathology

establishing coupling between ADDLs and synapse loss, neuron death, tau hyperphosphorylation and oxidative damage, the major pathological features of AD.

Given their putative involvement in AD pathogenesis, ADDLs provide significant opportunities for diagnostics and therapeutics. The large elevation of ADDLs in AD-CSF suggests that diagnostics may some day rely on ADDL-based assays. With respect to ADDL-based therapeutics, the most appealing approach is to target the pathogenic ADDLs themselves, not the cellular molecules that generate them. Given the soluble nature of the target, passive vaccines that specifically neutralize ADDLs are especially appealing and may obviate the side-effects and variable responses to active fibril-oriented vaccines observed in clinical trials (Ferrer et al. 2004). It also may be feasible to discover small molecules that could act as lead compounds for anti-oligomerization drugs (Chromy et al. 2003; De Felice et al. 2004; Wang et al. 2004b; Walsh et al. 2005). Given that ADDLs at least indirectly affect NMDA receptor signaling pathways and that the AD-drug Namenda is an NMDA receptor antagonist (Witt et al. 2004), prospects seem likely for new and more effective receptor-directed drugs to emerge from ADDL-based screens. Finally, because ADDLs are pathogenic ligands that attack specific synapses, the search for specific ADDL receptors as potential drug targets is actively underway.

Note This chapter is adapted from WL Klein, "Synaptic targeting by Aβ oligomers (ADDLs) as a basis for memory loss in early Alzheimer's disease". Alzheimer's and Dementia, 2:60–72, 2006, with permission from the Alzheimer's Association.

References

Allison DW, Chervin AS, Gelfand VI, Craig AM (2000) Postsynaptic scaffolds of excitatory and inhibitory synapses in hippocampal neurons: maintenance of core components independent of actin filaments and microtubules. J Neurosci 20:4545–4554

Alzheimer A (1906) Medical file for Auguste D, including admission report and interviews conducted bu author/doctor. In: Maurer K, Volk S, Gerbaldo H (1997) Auguste D and Alzheimer's disease. Lancet 349:1546–1549

Bigio EH, Lambert MP, Shaw P, Lacor PN, Viola KL, Klein WL (2005) Abeta oligomers in aging and Alzheimer disease. J Neuropathol Exp Neurol 64:440

Boutaud O, Montine TJ, Chang L, Klein WL, Oates JA (2006) PGH2-derived levuglandin adducts increase the neurotoxicity of amyloid β1-42. J Neurochem 96:917–923

Butler AE, Janson J, Soeller WC, Butler PC (2003) Increased beta-cell apoptosis prevents adaptive increase in beta-cell mass in mouse model of type 2 diabetes: evidence for role of islet amyloid formation rather than direct action of amyloid. Diabetes 52:2304–2314

Calon F, Lim GP, Yang F, Morihara T, Teter B, Ubeda O, Rostaing P, Triller A, Salem N, Jr., Ashe KH, Frautschy SA, Cole GM (2004) Docosahexaenoic acid protects from dendritic pathology in an Alzheimer's disease mouse model. Neuron 43:633–645

Carlisle HJ, Kennedy MB (2005) Spine architecture and synaptic plasticity. Trends Neurosci 28:182–187

Chang L, Bakhos L, Wang Z, Venton DL, Klein WL (2003) Femtomole immunodetection of synthetic and endogenous Amyloid-β oligomers and its application to Alzheimer's Disease drug candidate screening. J Mol Neurosci 20:305–313

Chen QS, Kagan BL, Hirakura Y, Xie CW (2000) Impairment of hippocampal long-term potentiation by Alzheimer amyloid beta-peptides. J Neurosci Res 60:65–72

Chin J, Palop JJ, Puolivali J, Massaro C, Bien-Ly N, Gerstein H, Scearce-Levie K, Masliah E, Mucke L (2005) Fyn kinase induces synaptic and cognitive impairments in a transgenic mouse model of Alzheimer's disease. J Neurosci 25:9694–9703

Chromy BA, Nowak RJ, Lambert MP, Viola KL, Chang L, Velasco PT, Jones BW, Fernandez SJ, Lacor PN, Horowitz P, Finch CE, Krafft GA, Klein WL (2003) Self-assembly of Aβ(1-42) into globular neurotoxins. Biochemistry 42:12749–12760

Conway KA, Lee SJ, Rochet JC, Ding TT, Williamson RE, Lansbury PT, Jr. (2000) Acceleration of oligomerization, not fibrillization, is a shared property of both alpha-synuclein mutations linked to early-onset Parkinson's disease: implications for pathogenesis and therapy. Proc Natl Acad Sci USA 97:571–576

Costello DA, O'Leary DM, Herron CE (2005) Agonists of peroxisome proliferator-activated receptor-gamma attenuate the Abeta-mediated impairment of LTP in the hippocampus in vitro. Neuropharmacol 49:359–366

De Felice FG, Vieira MN, Saraiva LM, Figueroa-Villar JD, Garcia-Abreu J, Liu R, Chang L, Klein WL, Ferreira ST (2004) Targeting the neurotoxic species in Alzheimer's disease: inhibitors of Abeta oligomerization. FASEB J 18:1366–1372

Dodart JC, Bales KR, Gannon KS, Greene SJ, DeMattos RB, Mathis C, DeLong CA, Wu S, Wu X, Holtzman DM, Paul SM (2002) Immunization reverses memory deficits without reducing brain Abeta burden in Alzheimer's disease model. Nature Neurosci 5:452–457

Ferrer I, Boada RM, Sanchez Guerra ML, Rey MJ, Costa-Jussa F (2004) Neuropathology and pathogenesis of encephalitis following amyloid-beta immunization in Alzheimer's disease. Brain Pathol 14:11–20

Fiala JC, Spacek J, Harris KM (2002) Dendritic spine pathology: cause or consequence of neurological disorders? Brain Res Brain Res Rev 39:29–54

Frackowiak J, Zoltowska A, Wisniewski HM (1994) Non-fibrillar beta-amyloid protein is associated with smooth muscle cells of vessel walls in Alzheimer disease. J Neuropathol Exp Neurol 53:637–645

Georganopoulou DG, Chang L, Nam JM, Thaxton CS, Mufson EJ, Klein WL, Mirkin CA (2005) Nanoparticle-based detection in cerebral spinal fluid of a soluble pathogenic biomarker for Alzheimer's disease. Proc Natl Acad Sci USA 102:2273–2276

Gong Y, Chang L, Viola KL, Lacor PN, Lambert MP, Finch CE, Krafft GA, Klein WL (2003) Alzheimer's disease-affected brain: Presence of oligomeric Aβ ligands (ADDLs) suggests a molecular basis for reversible memory loss. Proc Natl Acad Sci USA 100:10417–10422

Guzowski JF (2002) Insights into immediate-early gene function in hippocampal memory consolidation using antisense oligonucleotide and fluorescent imaging approaches. Hippocampus 12:86–104

Guzowski JF, Lyford GL, Stevenson GD, Houston FP, McGaugh JL, Worley PF, Barnes CA (2000) Inhibition of activity-dependent arc protein expression in the rat hippocampus impairs the maintenance of long-term potentiation and the consolidation of long-term memory. J Neurosci 20:3993–4001

Hardy J, Selkoe DJ (2002) The amyloid hypothesis of Alzheimer's disease: progress and problems on the road to therapeutics. Science 297:353–356

Hardy JA, Higgins GA (1992) Alzheimer's disease: the amyloid cascade hypothesis. Science 256:184–185

Hoshi M, Sato M, Matsumoto S, Noguchi A, Yasutake K, Yoshida N, Sato K (2003) Spherical aggregates of beta-amyloid (amylospheroid) show high neurotoxicity and activate tau protein kinase I/glycogen synthase kinase-3beta. Proc Natl Acad Sci USA 100:6370–6375

Huang X, Atwood CS, Moir RD, Hartshorn MA, Tanzi RE, Bush AI (2004) Trace metal contamination initiates the apparent auto-aggregation, amyloidosis, and oligomerization of Alzheimer's Abeta peptides. J Biol Inorg Chem 9:954–960

Jacobsen JS, Wu CC, Redwine JM, Comery TA, Arias R, Bowlby M, Martone R, Morrison JH, Pangalos MN, Reinhart PH, Bloom FE (2006) Early-onset behavioral and synaptic deficits in a mouse model of Alzheimer's disease. Proc Natl Acad Sci USA 103:5161–5166

Katzman R, Terry R, DeTeresa R, Brown T, Davies P, Fuld P, Renbing X, Peck A (1988) Clinical, pathological, and neurochemical changes in dementia: a subgroup with preserved mental status and numerous neocortical plaques. Ann Neurol 23:138–144

Kayed R, Head E, Thompson JL, McIntire TM, Milton SC, Cotman CW, Glabe CG (2003) Common structure of soluble amyloid oligomers implies common mechanism of pathogenesis. Science 300:486–489

Kelly MP, Deadwyler SA (2003) Experience-dependent regulation of the immediate-early gene arc differs across brain regions. J Neurosci 23:6443–6451

Klein WL (2001) Aβ toxicity in Alzheimer's Disease. In: Chesselet MF (ed) Molecular mechanisms of neurodegenerative diseases. Humana Press, Totowa, New Jersey, pp 1–49

Klein WL (2002) Abeta toxicity in Alzheimer's disease: globular oligomers (ADDLs) as new vaccine and drug targets. Neurochem Int 41:345

Klein WL (2005) Cytotoxic intermediates in the fibrillation pathway: Aβ oligomers in Alzheimer's disease as a case study. In: Uversky V (ed) Protein misfolding, aggregation, and conformational diseases. Kluwer Academic/Plenum Publishers, New York

Klunk WE, Engler H, Nordberg A, Wang Y, Blomqvist G, Holt DP, Bergstrom M, Savitcheva I, Huang GF, Estrada S, Ausen B, Debnath ML, Barletta J, Price JC, Sandell J, Lopresti BJ, Wall A, Koivisto P, Antoni G, Mathis CA, Langstrom B (2004) Imaging brain amyloid in Alzheimer's disease with Pittsburgh Compound-B. Ann Neurol 55:306–319

Klyubin I, Walsh DM, Cullen WK, Fadeeva JV, Anwyl R, Selkoe DJ, Rowan MJ (2004) Soluble Arctic amyloid beta protein inhibits hippocampal long-term potentiation in vivo. Eur J Neurosci 19:2839–2846

Klyubin I, Walsh DM, Lemere CA, Cullen WK, Shankar GM, Betts V, Spooner ET, Jiang L, Anwyl R, Selkoe DJ, Rowan MJ (2005) Amyloid beta protein immunotherapy neutralizes Abeta oligomers that disrupt synaptic plasticity in vivo. Nature Med 11:556–561

Kotilinek LA, Bacskai B, Westerman M, Kawarabayashi T, Younkin L, Hyman BT, Younkin S, Ashe KH (2002) Reversible memory loss in a mouse transgenic model of Alzheimer's disease. J Neurosci 22:6331–6335

Kuo YM, Emmerling MR, Vigo-Pelfrey C, Kasunic TC, Kirkpatrick JB, Murdoch GH, Ball MJ, Roher AE (1996) Water-soluble Abeta (N-40, N-42) oligomers in normal and Alzheimer disease brains. J Biol Chem 271:4077–4081

Lacor PN, Buniel MC, Chang L, Fernandez SJ, Gong Y, Viola KL, Lambert MP, Velasco PT, Bigio EH, Finch CE, Krafft GA, Klein WL (2004a) Synaptic targeting by Alzheimer's-related amyloid beta oligomers. J Neurosci 24:10191–10200

Lacor PN, Buniel MC, Klein WL (2004b) ADDLs (Aβ oligomers) alter structure and function of synaptic spines. 2004 Abstract Viewer/Itinerary Planner Washington, DC: Soc Neurosci Abstract No. 218.3

Lacor PN, Sanz-Clemente A, Viola KL, Klein WL (2005) Changes in NMDA receptor subunit 1 and 2B expression in ADDL-treated hippocampal neurons. 2005 Abstract Viewer/Itinerary Planner Washington, DC: Soc Neurosci Abstract No. 786.17

Lacor PN, Buniel MC, Furlow PW, Sanz-Clemente A, Velasco PT, Wood M, Viola KL, and Klein WL (2007) Abeta oligomer-induced aberrations in synapse composition, shape and density provide a molecular basis for loss of connectivity in Alzheimer's disease. J. Neurosci. in press

Lambert MP, Barlow AK, Chromy BA, Edwards C, Freed R, Liosatos M, Morgan TE, Rozovsky I, Trommer B, Viola KL, Wals P, Zhang C, Finch CE, Krafft GA, Klein WL (1998) Diffusible, nonfibrillar ligands derived from Abeta1-42 are potent central nervous system neurotoxins. Proc Natl Acad Sci USA 95:6448–6453

Lambert MP, Viola KL, Chromy BA, Chang L, Morgan TE, Yu J, Venton DL, Krafft GA, Finch CE, Klein WL (2001) Vaccination with soluble Abeta oligomers generates toxicity-neutralizing antibodies. J Neurochem 79:595–605

Lambert MP, Lacor PN, Chang L, Viola KL, Velasco PT, Richardson DK, Gong Y, Krafft GA, Klein WL (2003) ADDL-generated monoclonal antibodies target epitopes specific to Aβ oligomers. 2003 Abstract Viewer/Itinerary Planner Washington, DC: Soci Neurosci Abstract No. 527.16

Lambert MP, Velasco PT, Chang L, Viola KL, Fernandez S, Lacor PN, Khuon D, Gong Y, Bigio EH, Shaw P, De Felice FG, Krafft G, and Klein WL (2007) Monoclonal antibodies that target pathological assemblies of Abeta. J. Neurochem. 100:23–35

Lesne S, Koh MT, Kotilinek L, Kayed R, Glabe CG, Yang A, Gallagher M, Ashe KH (2006) A specific amyloid-beta protein assembly in the brain impairs memory. Nature 440:352–357

Lesort M, Jope RS, Johnson GV (1999) Insulin transiently increases tau phosphorylation: involvement of glycogen synthase kinase-3beta and Fyn tyrosine kinase. J Neurochem 72:576–584

Levine H, III (1995) Soluble multimeric Alzheimer beta(1-40) pre-amyloid complexes in dilute solution. Neurobiol Aging 16:755–764

Levine H, III (2004) Alzheimer's beta-peptide oligomer formation at physiologic concentrations. Anal Biochem 335:81–90

Masliah E, Miller A, Terry RD (1993) The synaptic organization of the neocortex in Alzheimer's disease. Med Hypotheses 41:334–340

Matsuzaki M, Ellis-Davies GC, Nemoto T, Miyashita Y, Iino M, Kasai H (2001) Dendritic spine geometry is critical for AMPA receptor expression in hippocampal CA1 pyramidal neurons. Nature Neurosci 4:1086–1092

McLean CA, Cherny RA, Fraser FW, Fuller SJ, Smith MJ, Beyreuther K, Bush AI, Masters CL (1999) Soluble pool of Abeta amyloid as a determinant of severity of neurodegeneration in Alzheimer's disease. Ann Neurol 46:860–866

Mesulam MM (1999) Neuroplasticity failure in Alzheimer's disease: bridging the gap between plaques and tangles. Neuron 24:521–529

Montalto MC, Agdeppa ED, Siclovan TM, Williams AC (2004) Composition and methods for non-invasive imaging of soluble beta-amyloid. New York/USA, Patent 10/747,715(US 2004/0223909 A1)

Mucke L, Masliah E, Yu GQ, Mallory M, Rockenstein EM, Tatsuno G, Hu K, Kholodenko D, Johnson-Wood K, McConlogue L (2000) High-level neuronal expression of abeta 1-42 in wild-type human amyloid protein precursor transgenic mice: synaptotoxicity without plaque formation. J Neurosci 20:4050–4058

Mukaetova-Ladinska EB, Garcia-Siera F, Hurt J, Gertz HJ, Xuereb JH, Hills R, Brayne C, Huppert FA, Paykel ES, McGee M, Jakes R, Honer WG, Harrington CR, Wischik CM (2000) Staging of cytoskeletal and beta-amyloid changes in human isocortex reveals biphasic synaptic protein response during progression of Alzheimer's disease. Am J Pathol 157:623–636

Nomura I, Kato N, Kita T, Takechi H (2005) Mechanism of impairment of long-term potentiation by amyloid beta is independent of NMDA receptors or voltage-dependent calcium channels in hippocampal CA1 pyramidal neurons. Neurosci Lett 391:1–6

Oda T, Pasinetti GM, Osterburg HH, Anderson C, Johnson SA, Finch CE (1994) Purification and characterization of brain clusterin. Biochem Biophys Res Commun 204:1131–1136

Oda T, Wals P, Osterburg HH, Johnson SA, Pasinetti GM, Morgan TE, Rozovsky I, Stine WB, Snyder SW, Holzman TF (1995) Clusterin (apoJ) alters the aggregation of amyloid beta-peptide (Aβ 1-42) and forms slowly sedimenting Aβ complexes that cause oxidative stress. Exp Neurol 136:22–31

Oddo S, Caccamo A, Shepherd JD, Murphy MP, Golde TE, Kayed R, Metherate R, Mattson MP, Akbari Y, LaFerla FM (2003) Triple-transgenic model of Alzheimer's disease with plaques and tangles: intracellular Abeta and synaptic dysfunction. Neuron 39:409–421

Oddo S, Caccamo A, Tran L, Lambert MP, Glabe CG, Klein WL, LaFerla FM (2006) Temporal profile of amyloid-beta (Abeta) oligomerization in an in vivo model of Alzheimer disease. A link between Abeta and tau pathology. J Biol Chem 281:1599–1604

Ohno M, Chang L, Tseng W, Oakley H, Citron M, Klein WL, Vassar R, Disterhoft JF (2005) Temporal memory deficits in Alzheimer's mouse models: Rescue by genetic deletion of BACE1 with reduced amyloid-β oligomers. Eur J Neurosci 23:251–260

Palop JJ, Chin J, Bien-Ly N, Massaro C, Yeung BZ, Yu GQ, Mucke L (2005) Vulnerability of dentate granule cells to disruption of arc expression in human amyloid precursor protein transgenic mice. J Neurosci 25:9686–9693

Puzzo D, Vitolo O, Trinchese F, Jacob JP, Palmeri A, Arancio O (2005) Amyloid-beta peptide inhibits activation of the nitric oxide/cGMP/cAMP-responsive element-binding protein pathway during hippocampal synaptic plasticity. J Neurosci 25:6887–6897

Reixach N, Deechongkit S, Jiang X, Kelly JW, Buxbaum JN (2004) Tissue damage in the amyloidoses: Transthyretin monomers and nonnative oligomers are the major cytotoxic species in tissue culture. Proc Natl Acad Sci USA 101:2817–2822

Scheff SW, Price DA (2003) Synaptic pathology in Alzheimer's disease: a review of ultrastructural studies. Neurobiol Aging 24:1029–1046

Serrano F, Klann E (2004) Reactive oxygen species and synaptic plasticity in the aging hippocampus. Ageing Res Rev 3:431–443

Sheng M, Lee SH (2001) AMPA receptor trafficking and the control of synaptic transmission. Cell 105:825–828

Snyder EM, Nong Y, Almeida CG, Paul S, Moran T, Choi EY, Nairn AC, Salter MW, Lombroso PJ, Gouras GK, Greengard P (2005) Regulation of NMDA receptor trafficking by amyloid-beta. Nature Neurosci 8:1051–1058

Steward O, Worley P (2002) Local synthesis of proteins at synaptic sites on dendrites: role in synaptic plasticity and memory consolidation? Neurobiol Learn Mem 78:508–527

Takahashi RH, Almeida CG, Kearney PF, Yu F, Lin MT, Milner TA, Gouras GK (2004) Oligomerization of Alzheimer's beta-amyloid within processes and synapses of cultured neurons and brain. J Neurosci 24:3592–3599

Tong L, Thornton PL, Balazs R, Cotman CW (2001) Beta - amyloid-(1-42) impairs activity-dependent cAMP-response element-binding protein signaling in neurons at concentrations in which cell survival Is not compromised. J Biol Chem 276:17301–17306

Trommer BL, Shah C, Yun SH, Gamkrelidze G, Pasternak ES, Blaine SW, Manelli A, Sullivan P, Pasternak JF, LaDu MJ (2005) ApoE isoform-specific effects on LTP: blockade by oligomeric amyloid-beta1-42. Neurobiol Dis 18:75–82

Vitolo OV, Sant'Angelo A, Costanzo V, Battaglia F, Arancio O, Shelanski M (2002) Amyloid beta-peptide inhibition of the PKA/CREB pathway and long-term potentiation: reversibility by drugs that enhance cAMP signaling. Proc Natl Acad Sci USA 99:13217–13221

Walsh DM, Klyubin I, Fadeeva JV, Cullen WK, Anwyl R, Wolfe MS, Rowan MJ, Selkoe DJ (2002) Naturally secreted oligomers of amyloid beta protein potently inhibit hippocampal long-term potentiation in vivo. Nature 416:535–539

Walsh DM, Townsend M, Podlisny MB, Shankar GM, Fadeeva JV, Agnaf OE, Hartley DM, Selkoe DJ (2005) Certain inhibitors of synthetic amyloid beta-peptide (Abeta) fibrillogenesis block oligomerization of natural Abeta and thereby rescue long-term potentiation. J Neurosci 25:2455–2462

Wang HW, Pasternak JF, Kuo H, Ristic H, Lambert MP, Chromy B, Viola KL, Klein WL, Stine WB, Krafft GA, Trommer BL (2002) Soluble oligomers of beta amyloid (1-42) inhibit long-term potentiation but not long-term depression in rat dentate gyrus. Brain Res 924:133–140

Wang Q, Walsh DM, Rowan MJ, Selkoe DJ, Anwyl R (2004a) Block of long-term potentiation by naturally secreted and synthetic amyloid beta-peptide in hippocampal slices is mediated via activation of the kinases c-Jun N-terminal kinase, cyclin-dependent kinase 5, and p38 mitogen-activated protein kinase as well as metabotropic glutamate receptor type 5. J Neurosci 24:3370–3378

Wang Z, Chang L, Klein WL, Thatcher GR, Venton DL (2004b) Per-6-substituted-per-6-deoxy beta-cyclodextrins inhibit the formation of beta-amyloid peptide derived soluble oligomers. J Med Chem 47:3329–3333

Westerman MA, Chang L, Frautschy S, Kotilinek L, Cole G, Klein W, Hsiao Ashe K (2002) Ibuprofen reverses memory loss in transgenic mice modeling Alzheimer's disease. Soc Neurosci Abstract 28:690.4

Witt A, Macdonald N, Kirkpatrick P (2004) Memantine hydrochloride. Nature Rev Drug Discov 3:109–110

Zhang C, Qiu HE, Krafft GA, Klein WL (1996) A beta peptide enhances focal adhesion kinase/Fyn association in a rat CNS nerve cell line. Neurosci Lett 211:187–190

Subject Index